Efficient Lighting Applications
and Case Studies

Efficient Lighting Applications and Case Studies

Scott C. Dunning, Ph.D., P.E., CEM
Albert Thumann, P.E., CEM

River Publishers

Routledge
Taylor & Francis Group
LONDON AND NEW YORK

Published 2020 by River Publishers
River Publishers
Alsbjergvej 10, 9260 Gistrup, Denmark
www.riverpublishers.com

Distributed exclusively by Routledge
4 Park Square, Milton Park, Abingdon, Oxon OX14 4RN
605 Third Avenue, New York, NY 10017, USA

First issued in paperback 2023

Library of Congress Cataloging-in-Publication Data

Dunning, Scott.
Efficient lighting applications and case studies / Scott C. Dunning, Ph.D., P.E.,
CEM; Albert Thumann, P.E., CEM.
pages cm
Includes index.
ISBN 0-88173-552-3 (alk. paper) -- ISBN 978-8-7702-2305-8 (electronic) (print) --
ISBN 978-1-4665-7154-9 (Taylor & Francis distribution : alk. paper) 1. Commercial
buildings--Lighting. 2. Electric lighting--Energy conservation. 3. Electric power-
-Conservation. I. Thumann, Albert. II. Title.

TH7975.C65D86 2012
621.32028'6--dc23

2012025798

Efficient Lighting Applications and Case Studies / by Scott C. Dunning and Albert Thumann
First published by Fairmont Press in 2013.

Routledge is an imprint of the Taylor & Francis Group, an informa business

Publisher's Note
The publisher has gone to great lengths to ensure the quality of this reprint but
points out that some imperfections in the original copies may be apparent.

ISBN 978-87-7022-918-0 (pbk)
ISBN 978-1-4665-7154-9 (hbk)
ISBN 978-8-7702-2305-8 (online)
ISBN 978-1-0031-5174-6 (ebook master)

While every effort is made to provide dependable information, the publisher,
authors, and editors cannot be held responsible for any errors or omissions.

Table of Contents

Acknowledgements

The information contained in this book has been obtained from a wide variety of authorities who are specialists in their respective fields. Appreciation is expressed to each of those authors who have contributed their expertise to this volume. Most of the chapters in this book were originally presented at the World Energy Engineering Congress sponsored by the Association of Energy Engineers.

Section I

Lighting Systems Design

Chapter 1

How to Design a Lighting System

With the increased concern for energy conservation in recent years, much attention has been focused on lighting energy consumption and methods for reducing it. Along with this concern for energy efficient lighting has come the realization that lighting has profound effects on worker productivity as well as important aesthetic qualities. This chapter presents an introduction to lighting design and some of the energy efficient techniques which can be utilized while maintaining the quality of illumination.

LAMP TYPES

There are seven different light sources that are popular today: incandescent, fluorescent, mercury vapor, metal halide, LED, high pressure sodium and low pressure sodium. All lamps except incandescent and LED are gas discharge lamps, meaning that light is created through the excitation of gases inside the lamp. All gas discharge lamps require a ballast. A ballast accomplishes the following functions:

1.   Limits the current flow.

2.   Provides a sufficiently high voltage to start the lamp.

3.   Provides the correct voltage to allow the arc discharge to stabilize.

4.   Provides power-factor correction to offset partially the coils' inductive reactance.

Lamp efficacy is determined by the amount of light, measured in lumens, produced for each watt of power the lamp requires. The lumens per watt (LPW) of the various light sources can vary considerably. Table 1-1 shows typical LPW ratings including power consumed by the ballast (ballast losses) where applicable.

3

Table 1-1

Lighting Comparison Chart		
Lighting Type	Efficacy (lumens/watt)	Lifetime(hours)
Incandescent		
Standard "A" bulb	10–17	750–2500
Energy-Saving Incandescent (or Halogen)	12–22	1,000–4,000
Reflector	12–19	2000–3000
Fluorescent		
Straight tube	30–110	7000–24,000
Compact fluorescent lamp (CFL)	50–70	10,000
Circline	40–50	12,000
High-Intensity Discharge		
Mercury vapor	25–60	16,000–24,000
Metal halide	70–115	5000–20,000
High-pressure sodium	50–140	16,000–24,000
Light-Emitting Diodes		
Cool White LEDs	60–92	25,000–50,000
Warm White LEDs	27–54	25,000–50,000
Low-Pressure Sodium	60–150	12,000–18,000

INCANDESCENT LAMPS

The incandescent lamp is one of the most common light sources. It is also the light source with the lowest efficacy (lumens per watt) and shortest life. This lamp is still popular, however, due to the simplicity with which it can be used and the low price of both the lamp and the fixture. Additionally, the lamp does not require a ballast to condition its power supply, light direction and brightness are easily controlled, and it produces light of high color quality.

The most common types of incandescent lamps are: the "A" or standard shaped lamp; the "PS" or pear-shaped lamp; the "R" or reflector

Table 1-1 (*Cont'd*)

Color Rendition Index (CRI)	Color Temperature (K)	Indoors/Outdoors
98–100 (excellent)	2700–2800 (warm)	Indoors/ outdoors
98–100 (excellent)	2900–3200 (warm to neutral)	Indoors/ outdoors
98–100 (excellent)	2800 (warm)	Indoors/ outdoors
50–90 (fair to good)	2700–6500 (warm to cold)	Indoors/ outdoors
65–88 (good)	2700–6500 (warm to cold)	Indoors/ outdoors
		Indoors
50 (poor to fair)	3200–7000 (warm to cold)	Outdoors
70 (fair)	3700 (cold)	Indoors/ outdoors
25 (poor)	2100 (warm)	Outdoors
70–90 (fair to good)	5000 (cold)	Indoors/ outdoors
70–90 (fair to good)	3300 (neutral)	Indoors/ outdoors
-44 (very poor)		Outdoors

Courtesy U.S. Dept. of Energy

lamp; the "PAR" or parabolic-aluminized reflector lamp and the tungsten-halogen (or quartz) lamp.

Light is produced in an incandescent lamp when the coiled tungsten filament is heated to incandescence (white light) by its resistance to a flow of electric current. The life of the lamp and its light output are determined by its filament temperature. The higher the temperature for a given lamp, the greater the efficacy and the shorter the life. The efficacy of incandescent lamps, however, does increase as the lamp wattage increase. This makes it possible to save on both energy and fixture costs whenever you can use one higher wattage lamp instead of two lower wattage lamps.

FLUORESCENT LAMPS

The fluorescent lamp is becoming the most common light source. It is easily distinguished by its tubular design-circular, straight or bent in a "U" shape. In operation, an electric arc is produced between two electrodes which can be several feet apart depending on the length of the tube. The ultraviolet light produced by the arc activates a phosphor coating on the inside wall of the tube, causing light to be produced.

Unlike the incandescent lamp, the fluorescent lamp requires a ballast to strike the electric arc in the tube initially and to maintain that arc. Proper ballast selection is important to optimum light output and lamp life.

Lamp sizes range from four watts to 215 watts. The efficacy (lumens per watt) of a lamp increases with lamp length. Reduced wattage fluorescent lamps and ballasts introduced in the last few years use from 10 percent to 20 percent less wattage than conventional fluorescent lamps.

Fluorescent lamps are available in a wide variety of colors but for most application the cool white, warm white and (newly introduced lite white) lamps produce acceptable color and high efficacy. Since fluorescent lamps are linear light sources with relatively low brightness as compared with point sources (incandescent and high intensity discharge lamps), they are suited for indoor application where lighting quality is important and ceiling heights are moderate.

Fluorescent lamp life is rated according to the number of operating hours per start, for example, 20,000 hours at three hours operation per start. The greater number of hours per start, the greater the lamp life. Because fluorescent lamp life ratings have increased, however, the number of times you turn a lamp on or off has become less important. As a general rule, if a space is to be unoccupied for more than a few minutes, the lamps should be turned off.

HIGH INTENSITY DISCHARGE LAMPS

High intensity discharge (HID) is the term used to designate four distinct types of lamps (mercury vapor, metal halide, high pressure sodium and low pressure sodium). Like fluorescent lamps they produce light by establishing an arc between two electrodes; however, in HID lamps the electrodes are only a few inches apart.

HID lamps require a few minutes (one to seven) to come up to full light output. Also, if power to the lamp is lost or turned off, the arc tube must cool before the arc can be restruck and light produced. Up to seven minutes (for mercury vapor lights) may be required.

MERCURY VAPOR LAMPS

The mercury vapor (MV) lamp produces light when electrical current passes through a small amount of mercury vapor. The lamp consists of two glass envelopes: an inner envelope in which the arc is struck, and an outer or protective envelope. The mercury vapor lamp, like the fluorescent lamp, requires a ballast designed for its specific use.

Although, used extensively in the past, mercury vapor lamps are not as popular as other HID sources today due to its relatively low efficacy. However, because of their low cost and long life (16,000 to 24,000 hours), mercury vapor lamps still find some applications.

The color rendering qualities of the mercury vapor lamp are not as good as those of incandescent lamps. A significant portion of the energy radiated is in the ultraviolet region resulting in a "bluish" light in the standard lamp. Through use of phosphor coatings on the inside of the outer envelope, some of the energy is converted to visible light resulting in better color rendition and use in indoor applications.

Mercury vapor lamp sizes range from 40 to 1,000 watts.

METAL HALIDE LAMPS

The metal halide (MH) lamp is very similar in construction to the mercury vapor lamp. The major difference is that the metal halide lamp contains various metal halide additives in addition to the mercury vapor. The efficacy of metal halide lamps is from 1.5 to 2 times that of mercury vapor lamps. The metal halide lamp produces a relatively "white" light, equal or superior to presently available mercury vapor lamps. The main disadvantage of the metal halide lamp is its relatively short life (7,500 to 20,000 hours).

Metal halide lamp sizes range from 175 to 1,500 watts. Ballasts designed specifically for metal halide lamps must be used.

HIGH PRESSURE SODIUM LAMPS

The high pressure sodium (HPS) lamp has the highest efficacy of all lamps normally used indoors. It produces light when electricity passes through a sodium vapor. This lamp also has two envelopes, the inner one being made of a poly crystalline aluminum in which the light-producing arc is struck. The outer envelope is protective, and may be either clear or coated. The light produced by this lamp is a "golden-white" color.

Although the HPS lamp first found its principal use in outdoor lighting, it now is a readily accepted light source indoors in industrial plants. It also is being used in many commercial and institutional applications as well.

HPS lamp size ranges from 35 to 1,000 watts. Ballasts designed specifically for high pressure sodium lamps must be used.

LOW PRESSURE SODIUM LAMPS

The low pressure sodium (LPS) is the most efficient light source presently available, providing up to 183 lumens per watt. The light in this lamp is produced by a U-shaped arc tube containing a sodium vapor. Its use indoors is severely restricted, however, because it has a monochromatic (yellow) light output. Consequently, most colors illuminated by this light source appear as tones of gray.

Low pressure sodium lamps range in size from 18 watts to 180 watts. Ballasts designed specifically for LPS must be used. The primary use of these lamps is street lighting as well as outdoor area and security lighting. Indoor applications such as warehouses are practical where color is not important.

LEDS

Luminaire Efficiency

In the previous section, it was seen that a lamp produces an amount of light (measured in lumens) which depends on the power consumed and the type of lamp. Equally important to the amount of light produced by a lamp, is the amount of light which is "usable" or provides illumination for the desired task. Luminaires, or lighting fixtures, are used to direct the light to a usable location, dependent on the specific requirements of the area to be lighted. Regardless of the luminaire type, some of the light is directed in non-usable directions, is absorbed by the luminaire itself or is absorbed by the walls, ceiling or floor of the room.

The coefficient of utilization, or CU, is a factor used to determine the efficiency of a fixture in delivering light for a specific application. The coefficient of utilization is determined as a ratio of light output from the luminaire that reaches the workplane to the light output of the lamps alone. Luminaire manufacturers provide CU data in their catalogs which are dependent on room size and shape, fixture mounting height and surface reflectances. Table 1-2 illustrates the form in which a vendor summarized the data used for determining the coefficient of utilization.

To determine the coefficient of utilization, the room cavity, ratio, wall reflectance, and effective ceiling cavity reflectance must be known.

Most data assume a 20% effective floor cavity reflectance. To determine the coefficient of utilization, the following steps are needed:

(a)   Estimate wall and ceiling reflectances.
(b)   Determine room cavity ratio.
(c)   Determine effective ceiling reflectance (pCC).

Step (a)
Typical reflectance values are shown in Table 1-3.

Steps (b) and (c)
Once the wall and ceiling reflectances are estimated it is necessary to analyze the room configuration to determine the effective reflectances. Any room is made up of a series of cavities which have effective reflectances with respect to each other and the work plane. Figure 1-1 indicates the basic cavities.

The space between fixture and ceiling is the ceiling cavity. The space between the work plane and the floor is the floor cavity. The space between the fixture and the work plane is the room cavity. To determine the cavity ratio use Figure 1-1 to define the cavity depth and then use Formula 1-1.

(Formula 1-1)   Cavity Ratio $= \dfrac{5 \times d \times (L+W)}{L \times W}$

Where
    d   = depth of the cavity as defined in Figure 1-1
    L   = Room (or area) length
    W = Room (or area) width

Table 1-2. Vendor Data for 175 Watt Mercury Vapor Lamp—Medium Spread Deflector

Coefficients of Utilization/Effective Floor Cavity Reflectance 20% (pFC)

% REFLECTANCE EFF. CEIL. (pCC)	WALL (pW)	ROOM CAVITY RATIO									
		1	2	3	4	5	6	7	8	9	10
80	50	0.854	0.779	0.711	0.647	0.591	0.539	0.490	0.446	0.407	0.355
	30	0.828	0.739	0.664	0.594	0.533	0.481	0.432	0.388	0.349	0.296
	10	0.805	0.705	0.626	0.552	0.491	0.440	0.392	0.347	0.309	0.258
70	50	0.832	0.761	0.698	0.635	0.578	0.530	0.483	0.438	0.401	0.349
	30	0.808	0.724	0.653	0.585	0.526	0.475	0.426	0.384	0.345	0.295
	10	0.786	0.695	0.618	0.546	0.486	0.434	0.387	0.344	0.308	0.256
50	50	0.788	0.725	0.669	0.610	0.558	0.511	0.466	0.424	0.388	0.339
	30	0.770	0.696	0.632	0.568	0.513	0.464	0.416	0.375	0.338	0.288
	10	0.754	0.670	0.602	0.534	0.478	0.428	0.382	0.339	0.303	0.253
30	50	0.750	0.694	0.642	0.587	0.539	0.495	0.450	0.412	0.377	0.329
	30	0.736	0.671	0.612	0.552	0.499	0.453	0.408	0.367	0.331	0.282
	10	0.722	0.649	0.586	0.523	0.469	0.421	0.375	0.335	0.299	0.249
10	50	0.716	0.665	0.618	0.566	0.521	0.479	0.438	0.399	0.366	0.319
	30	0.704	0.645	0.592	0.536	0.487	0.442	0.399	0.360	0.325	0.276
	10	0.693	0.628	0.571	0.511	0.460	0.413	0.370	0.330	0.294	0.245

Table 1-3. Typical Reflection Factors

COLOR	REFLECTION FACTOR
White and very light tints	.75
Medium blue-green, yellow or gray	.50
Dark gray, medium blue	.30
Dark blue, brown, dark green, and wood finishes	.10

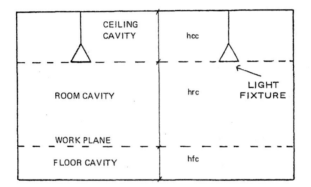

Figure 1-1. Cavity Configurations

To determine the effective ceiling or floor cavity reflectance, proceed in the same manner to define the ceiling or floor cavity ratio, then refer to Table 1-4 to find the corresponding effective ceiling or floor cavity reflectance.

SIM 1-1

For Process Plant No. 1, determine the coefficient of utilization for a room which measures 24' x 100'. The ceiling is 20' high and the fixture is mounted 4' from the ceiling. The tasks in the room are performed on work benches 3' above the floor. Use the data in Table 1-2.

Answer
Step (a)

Since no wall or ceiling reflectance data were given, assume a ceiling of .70 and wall of .5.

Table 1-4. Percent Effective Ceiling or Floor Cavity Reflectance for Various Reflectance Combinations

% Ceiling or Floor Reflectance →	90				80				70			50			30				10		
Ceiling or Floor Cavity Ratio ↓ / % Wall Reflectance →	90	70	50	30	80	70	50	30	70	50	30	70	50	30	65	50	30	10	50	30	10
0	90	90	90	90	80	80	80	80	70	70	70	50	50	50	30	30	30	30	10	10	10
0.1	90	89	88	87	79	79	78	78	69	69	68	50	49	48	30	30	29	29	10	10	10
0.2	89	88	86	85	79	78	77	76	68	67	66	49	49	47	30	29	29	28	10	10	9
0.3	89	87	85	83	78	77	75	74	68	66	64	49	48	46	30	29	28	27	11	10	9
0.4	88	86	83	81	78	76	74	72	67	65	63	48	47	45	30	28	27	26	11	10	9
0.5	88	85	81	78	77	75	73	70	66	64	61	48	46	44	29	28	27	25	11	10	9
0.6	88	84	80	76	77	75	71	68	65	62	59	47	46	43	29	28	26	25	11	10	9
0.7	88	83	78	74	76	74	70	66	65	61	58	47	45	42	29	28	26	24	11	10	8
0.8	87	82	77	73	75	73	69	65	64	60	56	47	44	41	29	27	25	23	11	10	8
0.9	87	81	76	71	75	72	68	63	63	59	55	46	43	40	29	27	25	22	11	9	8
1.0	86	80	74	69	74	71	66	61	63	58	53	46	43	39	29	27	24	22	11	9	8
1.1	86	79	73	67	74	71	65	60	62	57	52	46	42	38	29	26	24	21	11	9	8
1.2	86	78	72	65	73	70	64	58	61	56	50	45	41	37	29	26	23	20	12	9	7
1.3	85	78	70	64	73	69	63	57	61	55	49	45	41	36	29	26	23	20	12	9	7
1.4	85	77	69	62	72	68	62	55	60	54	48	45	40	35	28	25	22	19	12	9	7
1.5	85	76	68	61	72	68	61	54	59	53	47	44	40	34	28	25	22	18	12	9	7
1.6	85	75	66	59	71	67	60	53	59	52	45	44	39	33	28	25	21	18	12	9	7
1.7	84	74	65	58	71	66	59	52	58	51	44	43	39	32	28	25	21	17	12	9	7
1.8	84	73	64	56	70	65	58	50	57	50	43	43	38	32	28	25	21	17	12	9	6
1.9	84	73	63	55	70	65	57	49	57	49	42	43	37	31	28	25	20	16	12	9	6
2.0	83	72	62	53	69	64	56	48	56	48	41	43	37	30	28	24	20	16	12	9	6

(more)

2.1	2.2	2.3	2.4	2.5	2.6	2.7	2.8	2.9	3.0	3.1	3.2	3.3	3.4	3.5	3.6	3.7	3.8	3.9	4.0	4.1	4.2	4.3	4.4	4.5	4.6	4.7	4.8	4.9	5.0
6	6	6	6	6	5	5	5	5	5	5	5	5	5	5	5	4	4	4	4	4	4	4	4	4	4	4	4	4	4
9	9	9	9	9	9	9	9	9	8	8	8	8	8	8	8	8	8	8	8	8	8	8	8	8	8	8	8	8	8
13	13	13	13	13	13	13	13	13	13	13	13	13	13	13	13	13	13	13	13	13	13	13	13	14	14	14	14	14	14
16	15	15	14	14	13	13	13	12	12	12	11	11	11	11	10	10	10	10	9	9	9	9	8	8	8	8	8	7	7
20	19	19	19	18	18	18	18	17	17	17	16	16	16	16	15	15	15	15	15	14	14	14	14	14	14	13	13	13	13
24	24	24	24	23	23	23	23	23	22	22	22	22	22	22	21	21	21	21	21	21	20	20	20	20	20	20	19	19	19
28	28	28	28	27	27	27	27	27	27	27	27	27	27	26	26	26	26	26	26	26	26	26	26	25	25	25	25	25	25
29	29	28	28	27	26	26	25	25	24	24	23	23	22	22	21	21	21	20	20	20	19	19	19	19	18	18	18	18	17
36	36	35	35	34	34	33	33	33	32	32	31	31	31	30	30	30	29	29	29	28	28	28	27	27	27	26	26	26	26
43	42	42	42	41	41	41	41	40	40	40	40	39	39	39	39	38	38	38	38	37	37	37	37	37	36	36	36	36	36
40	39	38	37	36	35	34	33	33	32	31	30	30	29	29	28	27	27	26	26	25	25	25	24	24	24	23	23	23	22
47	46	46	45	44	43	43	42	41	40	40	39	39	38	38	37	37	36	36	35	35	34	34	34	33	33	33	32	32	32
56	55	54	54	53	53	52	52	51	51	50	50	49	49	48	48	48	47	47	46	46	46	45	45	45	44	44	44	44	43
47	45	44	43	42	41	40	39	38	38	37	36	35	34	33	33	32	31	30	30	29	29	28	28	27	26	26	25	25	25
55	54	53	52	51	50	49	48	48	47	46	45	44	44	43	42	42	41	40	40	39	39	38	38	37	37	36	36	35	35
63	63	62	61	61	60	60	59	58	58	57	57	56	56	55	54	54	53	53	52	52	51	51	51	50	50	49	49	49	48
69	68	68	67	67	66	66	66	65	65	64	64	64	63	63	62	62	62	61	61	60	60	60	59	59	59	58	58	58	57
52	51	50	48	47	46	45	44	43	42	41	40	39	38	37	36	35	35	34	33	32	32	31	30	30	29	29	28	28	27
61	60	59	58	57	56	55	54	53	52	51	50	49	48	48	47	46	45	45	44	43	43	42	41	41	40	40	39	38	38
71	70	69	68	68	67	66	66	65	64	64	63	62	62	61	60	60	59	59	58	57	57	56	56	55	55	54	54	53	53
83	83	82	82	82	82	82	81	81	81	80	80	80	80	79	79	79	79	78	78	78	78	78	77	77	77	77	76	76	76

Ceiling or Floor Cavity Ratio

Step (b)
    Assume 3′ working height.
    hrc = 20 − 4 − 3 = 13 (from Figure 1-1)
    From Formula 1-1, RCR = 3.4

Step (c)
    From Figure 1-1, hcc = 4
    From Formula 1-1, CCR= 1
    From Table 1-4, pCC = 58
    From Table 1-2, Coefficient of Utilization = 0.64 (interpolated)

Light Loss Factor
    The amount of light produced by a luminaire as determined by the lamp lumen output and the fixture coefficient of utilization is the initial value only. Over time the light reaching the task surface will depreciate due to two factors collectively known as the light loss factor (LLF).

    The light loss factor (LLF) takes into account that the lumen output of all lamps depreciates with time (LLD) and that the lumen output depreciates due to dirt build-up on the lamp and fixture (LDD). Formula 1-2 illustrates the relationship of these factors.

(Formula 1-2)   LLF = LDD x LLD

    To reduce the number of lamps required which in turn reduces energy consumption, it is necessary to increase the overall light loss factor. This is accomplished in several ways. One is to choose a luminaire which minimizes dust build-up. The second is to improve the maintenance program to replace lamps prior to burn-out. Thus if it is known that a group relamping program will be used at a given percentage of rated life, the appropriate lumen depreciation factor can be found. It may be decided to use a shorter relamping period in order to increase (LLD) even further.

    Figure 1-2 illustrates the effect of lumen depreciation and dirt build-up for a typical luminaire. Manufacturer's data should be consulted when estimating LLD and LDD for a luminaire.

Illumination Levels
    The amount of light that illuminates a surface is measured in lumens per square foot or footcandles. Table 1-3 shows selected illumination level ranges as recommended by the Illuminating Engineering society in the

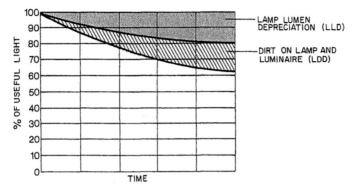

Figure 1-2. Light Output Reduction with Time

1981 Lighting Handbook. Note that these values are recommended for the performance of a specific task and that a room with various task areas would have various recommended illumination levels.

The values in Table 1-3 are intended as guidelines only. The age of the occupants, the inherent difficulty in viewing the object, the importance of speed and/or accuracy for visual performance and the reflectance of the task must be considered when applying these illumination ranges (see IES Handbook for further information).

The Lumen Method

Combining the concepts presented in the previous sections, we can use Formula 1-3 to determine the number of lamps to provide average, uniform lighting levels. This formula is known as the lumen method.

(Formula 1-3) $$N = \frac{E \times A}{Lu \times LLF \times CU}$$

where

N   is the number of lamps required
E   is the required illuminance in footcandles
A   is the area of the room in square feet
Lu is the lumen output of the lamp
LLF is the light loss factor which accounts for lamp lumen deprecia-tion and lamp (and fixture) dirt depreciation
CU is the coefficient of utilization

Table 1-3

Building Type	Space Type
Dormitories	Bedrooms
	Laundry rooms
Educational Buildings	Play room, nursery, classroom
	Lecture hall
	Computer practice rooms (menu driven)
Office buildings	Single offices
	Open plan offices
	Conference rooms
Educational buildings	Classrooms
	Classrooms for adult education
	Lecture hall
Hospitals	General ward lighting
	Simple examination
	Examination and treatment
Hotels and restaurants	Self-service restaurant, dining room
	Kitchen
	Buffet
Sport facilities	Sports halls
Wholesale and retail sales	Sales area
	Till area
Circulation areas	Corridor
	Stairs
	Restrooms
	Cloakrooms, washrooms, bathrooms, toilets
Industrial	Metal working/ welding
	Simple Assembly
	Difficult Assembly
	Exacting Assembly
Central Plant	Boiler house
	Machine Halls
	Side rooms, e.g. pump rooms, condenser rooms etc.
	Control rooms
Vehicle Construction/	Body work and assembly
Maintenance	Painting, spraying, polishing
	Painting, touch-up, inspection
Wood working and processing	Saw frame
	Work at joiner's bench, assembly
	Polishing, painting, fancy joinery

Table 1-3 (*Cont'd*)

Maintained Average Illuminance at working level (lux)	Measurement (working) Height (1 meter = 3.3 feet)
300	at 0 m
300	at 1 m
400	at 0 m
400	at 0.8 m
30	at 0.8 m
400	at 0.8 m
400	at 0.8 m
300	at 0.8 m
300	at 0.8 m
400	at 0.8 m
400	at 0.8 m
300	at 0.8 m
500	at 0.8 m
1000	at 0.8 m
100	at 0.8 m
500	at 0.8 m
100	at 0.8 m
300	at 0 m
500	at 0.8 m
500	at 0.8 m
50	at 0 m
50	at 0 m
300	at 0 m
300	at 0.8 m
300	at 1 m
300	at 1 m
1,000	
3,000-10,000	
50	at 0 m
300	
300	
500	
500	at 1 m
1000	
3,000-10,000	
300	at 1 m
300	
1000	

Note that for a specific area and level of illumination for the area, the only means that the lighting designer has for reducing the number of lamps (and consequently the power consumption) required is to use the highest values of Lu, CU and LLF.

SIM 1-2

For the situation described in Example Problem 1, determine the required number of fixtures to give an average, maintained footcandle level of 50. The light loss factor is estimated to be 0.7. The lamps are 175 watt mercury vapor with one lamp per fixture and an initial lumen output of 8,500.

Answer

$$\text{No. of fixtures} = \frac{\text{Area x Desired Maintained Footcandle}}{\text{Lumens x CU x LU}}$$

$$= \frac{24 \times 100 \times 50}{8,500 \times 0.64 \times 0.7} = 31.5 \text{ or } 32 \text{ fixtures}$$

The Point Method

The lumen method is useful in determining the average illumination in an area but sometimes it is desirable to know the illumination level due to one or more lighting fixtures upon a specified point within the area.

The point method (Formula 1-4) computes the level of illumination in footcandles by determining the contribution of a single light source in the area. For multiple light sources, Formula 1-4 must be used for each one and the results summed. Reflections from walls, ceilings and floors are not considered in this method, consequently it is especially useful for very large areas, outdoor lighting and areas where room surfaces are dark or dirty. Additionally, the formula holds true for point sources only. Caution must be exercised when using the point method for fluorescent sources or for luminaires with large reflectors. As a rule of thumb, if the maximum dimensions of the source are no more than one-fifth the distance to the point of calculation, the source will be considered a point source and the calculated illumination will be reasonably accurate.

$$\textbf{(Formula 1-4)} \quad \begin{array}{l} \text{Horizontal} \\ \text{Footcandles} \\ \text{on a Task} \end{array} = \frac{cp \times h}{d^2}$$

where
  cp   is the candle power at the desired angle Θ (see Figure 1-3) obtained from manufacturer's data (see Figure 1-4)
  h   is the height of the fixture above the horizontal plane of the task
  d   is the distance from the light source to the task

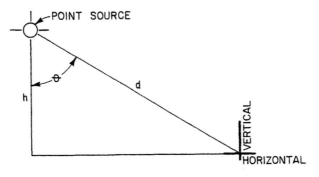

Figure 1-3. Point Source Method Orientations

A candlepower distribution curve (see Figure 1-4) shows the luminous intensity of a fixture (measured in candelas) for a range of angular orientations to the fixture. (0° is taken as directly beneath the fixture.)

SIM 1-3
What is the illumination on a surface due to a single 400 watt high pressure sodium light source represented by the data in Figure 1-4 which is 10′ in horizontal distance from the workplane? The vertical distance above the workplane (h) is 10′.

Answer
From trigonometric relationships, the angle Θ = arctangent 10/10 = 45°. From Figure 1-4 this gives a candlepower of 9500 candelas. d is found from trigonometric relationships to be h/cos Θ = 10/cos 45° = 14.1′. Therefore,

$$\text{Footcandles} = \frac{9500 \times 10}{(14.1)^3} = 33.9$$

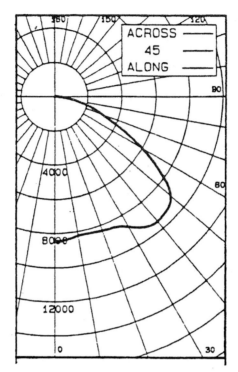

Figure 1-4. Candlepower Distribution for A 400 Watt HPS Low Bay Fixture

Fixture Layout

The fixture layout is dependent on the area. The initial layout should have equal spacing between lamps, rows and columns. The end fixture should be located at one-half the distance between fixtures. The maximum distance between fixtures usually should not exceed the mounting height unless the manufacturer specifies otherwise. Figure 1-5 illustrates a typical layout. If the fixture is fluorescent, it may be more practical to run the fixtures together. Since the fixtures are 4 feet or 8 feet long, a continuous wireway will be formed.

SIM 1-4

For SIM 1-2 design a lighting layout.

Answer

From SIM 1-2, thirty-two 175 watt mercury vapor lamps are required.

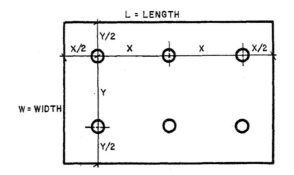

Figure 1-5. Typical Fixture Layout

		Rows	Columns	X Spacing	Y Spacing
Typical Combinations	(a)	4	8	12.5	6
	(b)	3	11	9	8
	(c)	2	16	6	12

(a)	(b)	(c)
$8X = 100$	$11X = 100$	$16X = 100$
$X = 12.5$	$X = 9$	$X = 6.2$
$4Y = 24$	$3Y = 24$	$2Y = 24$
$Y = 6$	$Y = 8$	$Y = 12$

Alternate (b) is recommended even though it requires one more fixture. It results in a good layout, illustrated following.

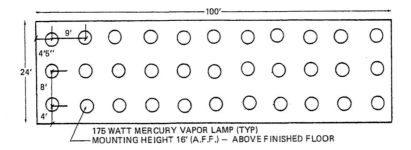

175 WATT MERCURY VAPOR LAMP (TYP)
MOUNTING HEIGHT 16' (A.F.F.) — ABOVE FINISHED FLOOR

CIRCUITING

Number of Lamps Per Circuit

A commonly used circuit loading is 1600 watts per lighting circuit breaker. This load includes fixture voltage and ballast loss. In SIM 1-4, assuming a ballast loss of 25 watts per fixture, a 20 amp circuit breaker, and #12 gauge wire, eight lamps could be fed from each circuit breaker. A single-phase circuit panel is illustrated in Figure 1-6. (Note: In practice ballast loss should be based on manufacturer's specifications.)

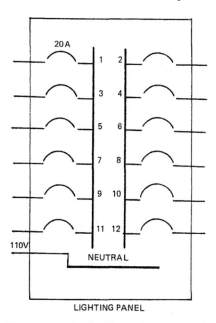

Figure 1-6. Single-Phase Circuit Panel

SIM 1-5

Next to each lamp place the panel designation and circuit number from which each lamp is fed; i.e., A-1, A-2, etc.

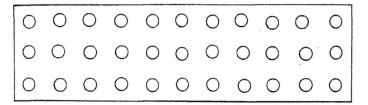

Answer

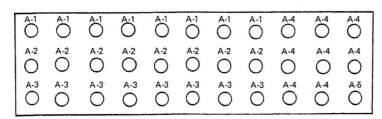

SIM 1-6

Designate a hot line from the circuit breaker with a small stroke and use a long stroke as a neutral; i.e., ⫴ 4 wires, 2 hot and 2 neutrals. The lamps are connected with conduit as shown below. Designate the hot and neutrals in each branch.

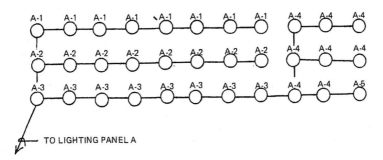

Hint-start wiring from the last fixture in the circuit.

Answer

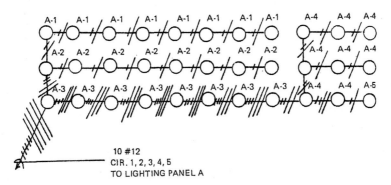

Note that only a single neutral wire is required for each 3 different phase wires.

POINTS ON LIGHTING DRAWINGS

- Choose a lighting drawing scale based on the area to be lighted and the detail required. Typical drawing scale: 1/8" equals one foot.

- Identify all symbols for lighting fixtures.

- Include circuit numbers on all lights.

- Include a note on fixture mounting height.

- Show "homerun" to lighting panels. "Homerun" indicates the number of wires and conduit size from the last outlet box.

- Use notes to simplify drawing. For example: All wires shall be 2 #12 in 3/4" conduit unless otherwise indicated. Remember the information put on a drawing or specification should be clear to insure proper illumination.

LIGHTING QUALITY

Illumination levels calculated by the lumen and point methods at best give only a "ballpark" estimate of the actual footcandle value to be realized in an installation. Many inaccuracies can be present including: differences between rated lamp lumen output and actual values; difficulty in predicting actual light loss factors; difficulty in predicting room surface reflectances; inaccurate CU information from a manufacturer; non-rectangular shaped rooms.

Precise illumination levels are not critically important, however. Of equal importance to lighting quantity is lighting quality. Very few people can perceive a difference of plus or minus ten footcandles, but poor quality lighting is readily apparent to anyone and greatly affects our ability to comfortably "see" a task.

Of the many factors affecting the quality of a lighting installation, glare has the greatest impact on our ability to comfortably perceive a task. Figure 1-7 shows the two types of glare normally encountered.

DIRECT GLARE

Direct glare is often caused by a light source in the midst of a dark surface. Direct glare also can be caused by light sources, including sunlight, in the worker's line of sight. A rating system has been developed for assessing direct glare called Visual Comfort Probability (VCP). The VCP takes into account fixture brightness at different angles of view, fixture size, room size, fixture mounting height, illumination level, and room surface reflectances.

Most manufacturers publish VCP tables for their fixtures. A VCP value of 70 or higher usually provides acceptable brightness for an office situation. Table 1-5 shows a typical VCP table.

INDIRECT GLARE

Indirect glare occurs when light is reflected off of a surface in the work area. When the light bounces off a task surface, details of the task surface become less distinct because contrast between the foreground and background,

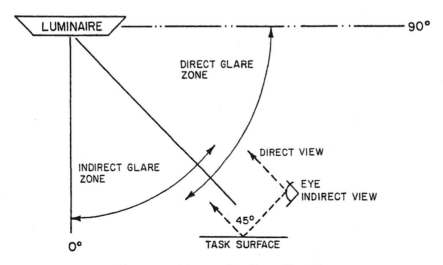

Figure 1-7. Direct and Indirect Glare Zones

such as the type on this page and the paper on which it is printed, is reduced. This is most easily visualized if a mirror is placed at the task surface and the image of a light fixture is seen at the normal viewing angle.

This form of indirect glare is called a veiling reflection because its effects are similar to those that would result were a thin veil placed between the worker's eyes and the task surface. Veiling reflections can be reduced by:

1) Orient fixtures (or work surface) so that the light produced is not in the indirect glare zone (generally to the side and slightly behind the work position gives the best results).

Table 1-5.

Room Size (in ft.)		Luminaires Lengthwise		Luminaires Crosswise	
		Ceiling Height (in feet)			
W	L	8.5	10.0	8.5	10.0
20 x	20	75	72	73	70
	30	75	72	73	70
	40	75	73	73	71
	60	75	73	72	71
30 x	20	78	75	77	73
	30	78	75	76	73
	40	77	75	75	72
	60	76	74	74	72
	80	76	74	73	72
40 x	20	81	78	80	77
	30	79	77	78	76
	40	78	77	76	75
	60	77	76	75	74
	80	77	75	74	73
	100	76	75	74	73
60 x	30	81	79	80	77
	40	79	78	78	76
	60	78	77	76	74
	80	77	76	75	74
	100	77	76	75	73
100 x	40	81	80	80	78
	60	80	78	78	76
	80	78	77	77	75
	100	78	76	76	74

**Wall Reflectance, 50%
Ceiling Cavity Reflectance, 80%
Floor Cavity Reflectance, 20%
Work Plane Illumination, 100fc

2) Select fixtures which direct the light above the worst veiling angles (generally 30° or greater). These fixtures have "batwing" distribution patterns such as shown in Figure 1-4. Note, that in selecting fixtures to minimize indirect glare, care must be taken not to select fixtures that are a source of excessive direct glare.

ENERGY CONSERVATION CONSIDERATIONS

Recent concern for energy conservation has focused attention on lighting as an area for potential savings since it can account for 25% of total energy use in an office building. This attention has resulted in many new products which greatly decrease the amount of energy needed for lighting. Unfortunately, blanket application of energy conserving techniques has also resulted in some poorly lit applications which save energy at the expense of worker productivity.

LAMP/LUMINAIRE EFFICIENCIES

As noted in the previous sections, there are wide variations in the efficacies of light sources. By selecting the most efficient light source within the color and room configuration constraints, significant energy savings are possible. Additionally, a new generation of "energy efficient" fluorescent lamps and ballasts are available which offer 5-20% savings over their standard counterparts.

Also, as shown in the previous sections, the choice of luminaire can have a great impact on the energy used for lighting since it determines how much light reaches the task.

NON-UNIFORM LIGHTING

The lumen method presented previously is useful for calculating average uniform illumination for an area, but illumination levels presented in Table 1-3 are for specific tasks in an area. By tailoring illumination levels to the various tasks in an area, significant energy savings are possible.

For example, if 30% of an office area is comprised of desks at which people will be reading material written in pencil, the recommended illumination level is from 50 to 100 footcandles depending on the application. In the remaining 70% of the office, however, if the area is used for general passageways or as a lobby area, the required illumination level is only 10 to 20 footcandles. By directing high levels of illumination only to the task areas, significant energy savings are possible. (Note, that it is generally recommended to limit the ratio of task levels to non-task levels to 3 to 1 to minimize fatigue caused by excessive contrast.)

This technique is referred to as non-uniform or task ambient lighting. It can be accomplished by positioning fixtures over the task location or by providing a small fixture which is mounted on the desk or machine tool, for example, to provide localized lighting.

GROUP RELAMPING

As seen in Figure 1-2 light output of a fixture depreciates over time due to lamp aging and dirt accumulation. If a minimum footcandle level is desired, initial foot candles must be as much as 40% higher than that desired. Consequently, many more fixtures and consequent power consumption is required to compensate for these depreciation factors.

If a systematic program is initiated, however, to periodically clean the fixtures and relamp before the end of rated life, the number of fixtures can be reduced while maintaining the desired illumination level. This technique is known as group relamping and lighting maintenance. It has the effect of raising the light loss factor (LLF) in Formula 1-3.

LEVEL CONTROLS

Areas with daylight available through windows and skylights can achieve significant energy savings by reducing the lighting system output to maintain a desired illumination level. This can be accomplished by either turning the fixture off, by reducing its output by switching off some of the lamps in the fixture or by using special dimming circuitry.

Additionally, level controls can reduce lighting system energy consumption during "non-production" times when lights are needed. For example, an office which required 70 footcandles during business hours only required 20 footcandles for cleaning at night. By reducing the light levels to 20 footcandles during the cleaning periods, significant energy savings are possible.

Also, level controls (specifically dimming controls) can be used to compensate for light depreciation factors thereby providing required footcandle levels with the minimum possible power consumption.

ON/OFF CONTROLS

One of the simplest and most effective means of controlling lighting energy consumption is by turning off the lights when not needed. To effectively accomplish this may require the addition of switches in each office, grouping of lights into "zones" of usage types, the use of occupancy sensors (either ultrasonic or infrared) to detect when occupants are present and/or the use of an energy management system to automatically schedule lighting operation.

JOB SIMULATION-SUMMARY PROBLEM

SIM 1-7
(a)   The Ajax Plant contains a workshop area with an area of 20' x 18'6."

For this area compute the number of lamps required, the space between fixtures, and the circuit layout. Use two 40-watt fluorescent lamps per fixture, 2900 lumens per lamp, light loss factor = .7, 110-volt lighting system, 20-watt ballast loss per fixture, and a fixture length of 2' x 4'. Use luminaire data of Table 1-6, ceiling height 20' and a desired footcandle level of 40.

Analysis

The area of the workshop is 18'6" x 20'.

Assume   hfc = 3

            hcc = 3

Therefore, hrc = 14

Assume   70% ceiling reflectance

            50% wall reflectance

The room cavity ratio is 7 and the effective ceiling cavity ratio pcc = 53. Thus C.U. = .55.

$$\text{No. of Fixtures} \quad \frac{20' \times 18\text{-}1/2 \times 40}{2 \times 2900 \times .55 \times .70} = 7$$

Table 1-6. Coefficient of Utilization 20% Effective Floor Cavity Reflectance

Effective Ceiling Cavity Reflectance	80%			50%		
Wall Reflectance	50	30	10	50	30	10
RCR						
10	.33	.26	.22	.31	.26	.22
9	.43	.35	.27	.40	.35	.29
8	.58	.42	.35	.48	.42	.36
7	.58	.50	.42	.55	.48	.42
6	.64	.57	.49	.61	.54	.47
5	.72	.65	.59	.65	.60	.56
4	.77	.71	.64	.71	.65	.60
3	.82	.76	.70	.74	.69	.63
2	.87	.82	.77	.78	.74	.70
1	.91	.87	.83	.81	.78	.75
Spacing not to exceed 1 X Mounting Height						

Each 20 amp lighting circuit can provide power for up to 16 fixtures.

Layout Spacing

$3x = 20$

$x = 6.6$

$3y = 18W$

$y = 6'2''$

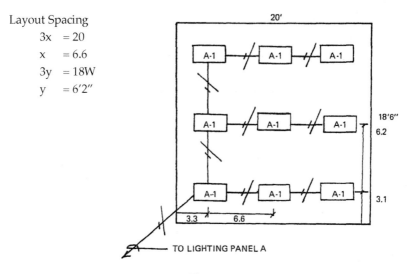

Figure 1-8.

Note: With emphasis on energy conservation, a lighting layout using 6 fixtures may be preferable.

SUMMARY

Energy conservation is influencing lighting design. Increased emphasis is being placed on minimizing lighting energy use by using lamps and luminaries which have high lumen outputs and coefficients of utilization. Today's lighting systems incorporate switching and automatic control devices to make it easy to turn off lights when they are not required. Lighting systems need to be analyzed on a first and operating cost basis to insure that the increasing energy costs are taken into account.

Lighting system design must not only consider the quantity of illumination but also the quality of illumination. The choice of a luminaire and its location play an important part in comfortably perceiving a task. An awareness of the importance of quality lighting can result in a visual environment which is productive as well as energy efficient.

Chapter 2

Selection Criteria for Lighting Energy Management

Roger L. Knott, P.E.

Today there are many tools available to the designer and facility manager to aid in lighting energy management. Even before the energy concerns became critical in the 1970s, the lighting industry had made substantial progress in improved lamp efficacy* and higher lighting system efficiency.

LIGHT SOURCES

Figure 2-1 indicates the general lamp efficiency ranges for the generic families of lamps most commonly used for both general and supplementary lighting systems. Each of these sources is discussed briefly here. It is important to realize that in the case of fluorescent and high intensity discharge lamps, the figures quoted for "lamp efficacy" are for the lamp only and do not include the associated ballast losses. To obtain the total system efficiency, ballast input watts must be used rather than lamp watts to obtain an overall system lumen per watt figure. This will be discussed in more detail in a later section.

Incandescent lamps have the lowest range of lamp efficacies of the commonly used lamps. This would lead to the accepted conclusion that incandescent lamps should, generally, not be used for large area, general lighting systems where a more efficient source could serve satisfactorily. However, this does not mean that incandescent lamps should never be used. There are many applications where the size, convenience, easy control, color rendering, and relatively low cost of incandescent lamps are suitable for a specific application.

---

*As used in this chapter, this term refers to "luminous efficacy of a source of light" which is defined as the quotient of the total luminous flux emitted divided by the total lamp power input. It is expressed in lumens per watt. See also IES Lighting Handbook, 1984 Reference Volume, The Illuminating Engineering Society of North America, New York, N.Y. 10017.

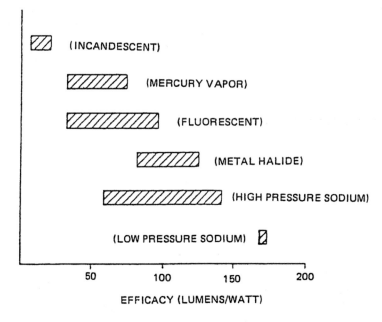

Figure 2-1. General Service Lamp Efficacy

General service incandescent lamps do not have good lumen maintenance throughout their lifetime. This is the result of the tungsten being evaporated off the filament during heating and being deposited on the bulb wall, thus darkening the bulb and reducing the lamp lumen output. Tungsten halogen (quartz) lamps do not suffer from this problem because they use a halogen regenerative cycle so that the tungsten driven off the filament is redeposited on the filament rather than the bulb wall. Therefore, the tungsten-halogen lamps retain lumen outputs in excess of 95 percent of initial values throughout their lifetime.

Mercury vapor lamps find limited use in today's lighting systems because fluorescent and other high intensity discharge (HID) sources have passed them in both lamp efficacy and system efficiency. Typical ratings for mercury vapor lamps range from about 30 to 70 lumens per watt. The primary advantages of mercury lamps are a good range of color, availability in sizes as low as 30 watts, long life and relatively low cost. However, fluorescent systems are available today which can do many of the jobs mercury used to do and they do it more efficiently. There are still places for mercury vapor lamps in lighting system design, but they are becoming fewer as technology advances in fluorescent and higher efficacy HID sources.

Fluorescent lamps have made dramatic advances in the last 10 years. From the introduction of reduced wattage lamps immediately following the Arab oil embargo of the mid 1970s, to the marketing of several styles of low wattage, compact lamps recently, there has been a steady parade of new products. Lamp efficacy now ranges from about 30 lumens per watt to near 90 lumens per watt. The range of colors is more complete than mercury vapor and lamp manufacturers have recently made significant progress in developing fluorescent and metal halide lamps which have much more consistent color rendering properties allowing greater flexibility in mixing these two sources without creating disturbing color mismatches. The recent compact fluorescent lamps open up a whole new market for fluorescent sources. These lamps permit design of much smaller luminaries which can compete with incandescent and mercury vapor in the low cost, square or round fixture market which the incandescent and mercury sources have dominated for so long. While generally good, lumen maintenance throughout the lamp lifetime is a problem for some fluorescent lamp types.

Metal halide lamps fall into a lamp efficacy range of approximately 75-125 lumens per watt. This makes them more energy efficient than mercury vapor but somewhat less so than high pressure sodium. Metal-halide lamps generally have fairly good color rendering qualities. While this lamp displays some very desirable qualities, it also has some distinct drawbacks including relatively short life for an HID lamp, long restrike time to restart after the lamp has been shut off (about 15-20 minutes at 70°F) and a pronounced tendency to shift colors as the lamp ages. In spite of the drawbacks, this source deserves serious consideration and is used very successfully in many applications.

High pressure sodium lamps introduced a new era of extremely high efficacy (60-140 lumens/watt) in a lamp which operates in fixtures having construction very similar to those used for mercury vapor and metal halide. When first introduced, this lamp suffered from ballast problems. These have now been resolved and luminaries employing high quality lamps and ballasts provide very satisfactory service. The 24,000-hour lamp life, good lumen maintenance and high efficacy of these lamps make them ideal sources for industrial and outdoor applications where discrimination of a range of colors is not critical.

The lamp's primary drawback is the rendering of some colors. The lamp produces a high percentage of light in the yellow range of the spectrum. This tends to accentuate colors in the yellow region. Rendering of reds and greens show a pronounced color shift. This can be compensated for in the selection of the finishes for the surrounding areas and, if properly done, the results can be very pleasing. In areas where color selection, matching and discrimination are necessary, high pressure sodium should not be used as the only source of light. It is possible to gain quite satisfactory color rendering by mixing high pressure sodium and metal halide in the proper proportions. Since both sources have relatively high efficacies, there is not a significant loss in energy efficiency by making this compromise.

High pressure sodium has been used quite extensively in outdoor applications for roadway, parking and facade or security lighting. This source will yield a high efficiency system; however, it should be used only with the knowledge that foliage and landscaping colors will be severely distorted where high pressure sodium is the only, or predominant, illuminant. Used as a parking lot source, there may be some difficulty in identification of vehicle colors in the lot. It is necessary for the designer or owner to determine the extent of this problem and what steps might be taken to alleviate it.

Recently lamp manufacturers have introduced high pressure sodium lamps with improved color rendering qualities. However, as with most things in this world, the improvement in color rendering was not gained without cost-the efficacy of the color improved lamps is somewhat lower, approximately 90 lumens per watt.

Low pressure sodium lamps provide the highest efficacy of any of the sources for general lighting with values ranging up to 180 lumens per watt. Low pressure sodium produces an almost pure yellow light with very high efficacy and renders all colors gray except yellow or near yellow. The effect of this is there can be no color discrimination under low pressure sodium lighting and it is suitable for use in a very limited number of applications. It is an acceptable source for warehouse lighting where it is only necessary to read labels but not to choose items by color. This source has application for either indoor or outdoor safety or security lighting, again as long as color rendering is not important.

In addition to these primary sources, there are a number of retrofit lamps which allow use of higher efficacy sources in the sockets of existing fixtures. Therefore, metal halide or high pressure sodium lamps can be retrofitted into mercury vapor fixtures or self ballasted mercury lamps can replace incandescent lamps. These lamps all make some compromises in operating characteristics, life and/or efficacy.

Figure 2-2 presents data on the efficacy of each of the major lamp types in relation to the wattage rating of the lamps. Without exception, the efficacy of the lamp increases as the lamp wattage rating increases.

The lamp efficacies discussed here have been based on the lumen output of a new lamp after 100 hours of operation or the "initial lumens." Like people, not all lamps age in the same way. Some lamp types, such as lightly loaded fluorescent and high pressure sodium as shown in Figures 2-3 and 2-4, hold up well and maintain their lumen output at a relatively high level until they are into, or past, middle age. Others, as represented by heavily loaded fluorescent, mercury vapor and metal halide, decay rapidly during their early years and then coast along at a relatively lower lumen output throughout most of their useful life. These factors must be considered when evaluating the various sources for overall energy efficiency.

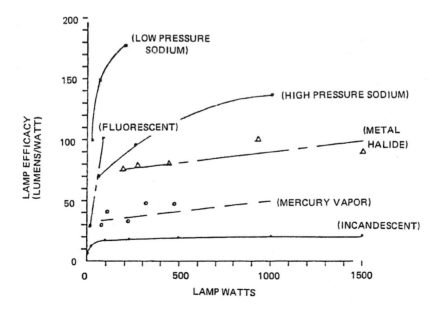

Figure 2-2. Lamp Efficacy (Does Not Include Ballast Losses)

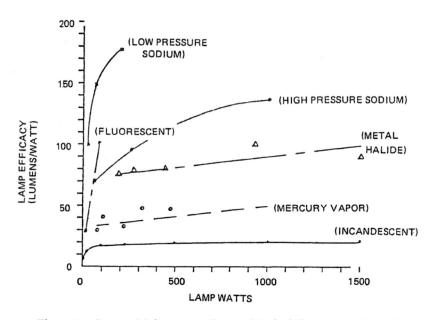

Figure 2-3. Lumen Maintenance Curves (Typical Fluorescent Lamps)

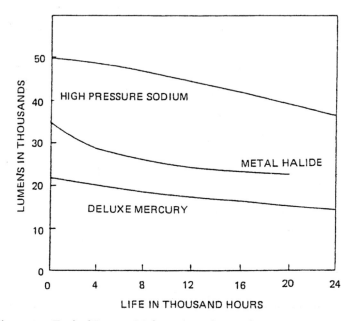

Figure 2-4. Typical Lumen Maintenance Curves (400 Watt H.I.D. Lamps)

LUMINAIRE SELECTION

The luminaire selected can have a significant impact on the energy efficacy of the system as a whole. By selecting a luminaire which generates a distribution of light that results in a high quality and reasonable quantity illuminance, workers will be able to perform their tasks more efficiently. A luminaire which is applied in such a way that direct glare, reflected glare and veiling reflections are kept to a minimum will increase worker productivity.

An important consideration in the efficiency of a lighting system is the coefficient of utilization (CU). This is a measure of the efficiency with which the luminaire distributes the lumens generated by the light source to the space. A fixture with a high CU will generally produce higher foot candle levels than one with a lower CU. However, the advantages of a high CU might be lost if the distribution is not carefully controlled. It is often found that luminaires with lower CU's result in lighting systems of superior quality and allow better task visibility and productivity with lower illuminances because of the improved quality. To date there has been no successful method developed which can effectively correlate productivity mathematically with lighting quality or quantity.

With all discharge lamps, fluorescent and HID, it is necessary to employ a ballast to provide starting voltage and control operating current. Advances have been made in recent years which have reduced the ballast losses. Many low loss ballasts are available for fluorescent lamps to provide higher operating efficiencies. Some provide maximum power savings with reduced lamp lumen output and others maintain full light output while reducing input watts.

Some energy efficient systems on the market use three-lamp, 2 ft x 4 ft fluorescent luminaires which provide illuminance equal to four-lamp luminaires employing standard lamps and ballasts. These three-lamp fixtures may reduce the power consumed by 30 percent compared to the standard four-lamp fixtures.

While there have been improvements in HID ballasts over the years, the selection available to the user is, generally, more restricted than in the case of fluorescent.

Finally, the maintainability of the luminaire must be considered as it relates to power and energy efficiency. In industrial facilities particularly, a fixture which is easily maintained is important. Maintenance, generally speaking, is not routinely performed on the lighting system.

If a luminaire is cleaned at all, it generally is cleaned only when it is relamped. With the introduction of longer life lamps, many up to 24,000 hours rated life, relamping may not occur for several years. Without maintenance, the dirt which collects on the luminaire and the lamp can greatly reduce the lumen output of the system.

Some luminaire features should be considered which will improve the maintainability of the system:

Ventilated luminaires which allow, in fact encourage, the passage of air through the lamp chamber help reduce dirt buildup on the luminaire surfaces. This constant flow of air in the bottom and out the top of the reflector tends to carry airborne dirt and contaminants through the fixture without allowing them to settle out on the lamp or reflector. Without this action, oil or grease can collect in greater quantities and allow other dirt particles to stick to the surfaces.

Reflector surfaces can also be effective in reducing the dirt which will adhere. At least one manufacturer realized the importance of a smooth, hard surface many years ago and made the inner surface of their luminaire a smooth glass with the light control prisms on the outer surface. Since the prisms were designed to reflect the light back through the fixture, dirt collecting on the outside, rear surface of the refractor has no impact on the efficiency of the luminaire. Today there are many suitable treatments available, such as Alzak™, Alglas™ and applied silicone surfaces which provide stable, hard, and easily cleaned reflector surfaces and contribute to the continued efficiency, both lighting and energy, of the system.

Closures can be provided in particularly dirty environments to stop the dirt and contaminants before they reach the lamp and reflector. Closures can be particularly important in damp, wet or extremely dirty atmospheres where the quantities of contaminant could pose a problem to the continued safe operation of the system, such as situations where steam or water vapor could short out lamp sockets or acid vapors could attack reflector surfaces. Where conditions warrant, the closure should be provided with a gasket.

Breathers can be added to the luminaire to further control the contaminants introduced in enclosed fixtures. Even though a gasketed closure is provided over the bottom of the reflector, as the luminaire alternately heats and cools from being turned on and off, it breathes as the air inside expands and contracts. If this air is drawn through the gasketed perimeter of the closure, it may carry significant quantities of airborne contaminants with it. By adding a breather in the reflector assembly, much of the air passes in and out through the breather and at least a portion of the contaminants will be trapped, thereby reducing the buildup on the luminaire.

SPACE CHARACTERISTICS

Room surfaces can contribute to lighting system power and energy efficiency. This is reasonable considering that much of the illuminance in the space is the result of light interreflected from the room surfaces.

Higher reflectance room surfaces reduce the number of luminaires needed by developing a higher luminaire CU (see Table 2-1). Since the CU affects the Zonal Cavity Calculation for number of luminaires required indirectly, the higher CU results in fewer fixtures required to provide the desired illuminance as can be seen using the following formula for number of fixtures required:

$$N = \frac{FC \times A}{L/L \times CU \times LLF}$$

Where

N	=	Number of luminaires required
FC	=	Illuminance in footcandles (or lux)
A	=	Area in square feet (or square meters)
L/L	=	Lumens per luminaire
CU	=	Coefficient of Utilization
LLF	=	Product of all of the light loss factors

Fewer luminaires will, of course, require less power and thereby, consume less energy.

The selection of room surface reflectances requires some care. Low reflectances, as noted above, result in lowered CU and an increased number of luminaires to provide the desired illuminance. However, reflectances which are too high can result in glare and/or excessive surface luminance ratios within the space. Glossy finishes should be avoided as they are potential glare sources.

When considering room finishes, such consideration should not be limited to the room surfaces. The equipment located within the space has

Table 2-1. Typical Photometrics

COEFFICIENTS OF UTILIZATION   ZONAL CAVITY

fc = 20%

cc →	80%				70%				50%			30%		
w \ cc(w%)	70%	50%	30%	10%	70%	50%	30%	10%	50%	30%	10%	50%	30%	10%
1	76	73	71	68	74	72	69	67	69	67	65	66	65	63
2	70	65	61	58	68	64	61	57	62	59	56	60	57	55
3	65	59	54	50	63	58	53	50	56	52	49	54	51	48
4	60	53	48	44	59	52	47	43	50	46	43	49	45	42
5	55	47	42	38	54	47	42	38	45	41	37	44	40	37
6	51	43	37	33	50	42	37	33	40	36	33	40	36	33
7	48	39	33	29	46	38	33	29	37	33	29	36	32	29
8	44	35	29	26	43	35	29	26	34	29	25	33	28	25
9	40	31	26	22	39	31	26	22	30	25	22	30	25	22
10	38	29	23	20	37	28	23	20	28	23	20	27	23	20

reflecting surfaces which should be included in considering colors, reflectances and surface finish since they will have an effect upon the lighting system.

Daylighting can be used effectively to reduce electrical lighting energy by the proper use of windows, clerestory, monitor roofing or skylights in industrial or office areas. It is usually necessary to provide luminaires throughout the space for necessary lighting during periods when adequate daylight is not available. However, by selective controls, this part of the lighting system can be dimmed or turned off when daylight is available, reducing the energy consumed even though it may not reduce the connected load.

When using daylighting, careful consideration must be given to proper shading and control devices to avoid annoying glare and brightness patterns. It may be possible to increase the effectiveness of daylighting by locating those tasks which can make effective use of daylight in areas where such light is available. Any use of daylighting within the building should be carefully coordinated with the building heating and cooling requirements so that it will be analyzed with the total building energy consumption.

MAINTENANCE

Reference to the formula for the zonal cavity method of lighting calculations in the previous section shows there is a term referred to as LLF—light loss factors. Generally speaking, these factors can be grouped into eight categories,[1] as:

1.   Luminaire ambient temperature
2.   Voltage to luminaire
3.   Ballast factor
4.   Luminaire surface depreciation
5.   Room surface dirt depreciation
6.   Lamp lumen depreciation
7.   Lamp burnout factor
8.   Luminaire dirt depreciation

The first four of these factors can be characterized as "unrecoverable" since there is little that a maintenance program can do to improve them. It may be possible to achieve some improvement in the ambient temperature and voltage which affect the luminaire but it is quite likely that any attempt to significantly affect these factors would not prove to be cost effective. The better solution, if these problems are serious in a given application, would be to use an alternate light source which would be less affected by the particular factor.

Four factors do represent areas which are considered recoverable because they can usually be completely restored by the proper maintenance procedures.

Room surface dirt depreciation (RSDD) is dependent upon the space dirt condition and the procedures used to maintain room surfaces. It was shown previously that the reflectance of the space surfaces has a significant impact on the luminaire CU. The associated surface dirt depreciation factor can also impact the number of luminaires required by the Zonal Cavity Calculation. The conditions affecting the RSDD factor are luminaire distribution type, expected space dirt conditions and the size of the space.

Lamp lumen depreciation (LLD) occurs in all lamps as they age. The lamp lumen depreciation factor is a direct result of the type of lamp and the replacement program followed by the facility. If lamps are spot replaced after they have failed, the LLD factor used for calculating illuminance will be low. However, if there is a group lamping program at the facility, a relamping period can be established to replace lamps at some appropriate time prior to the end of their lifetime. This will result in a higher LLD and can also result in fewer luminaires installed in the space.

Lamp burnout factor (LBO) is directly associated with the maintenance procedures to be followed in the given facility. This factor does have a relationship with the lamp lumen depreciation (LLD) factor. If the facility chooses to use spot relamping and relamps as each failure occurs, the LLD will be low, because lamps burn to end of life and operate through the most inefficient portion of their life cycle. However, under these conditions, the LBO will be high because lamps will be replaced as they burn out and no sockets will be left with burned-out lamps. On the other hand, if a group relamp method is chosen, lamps will be replaced before their end of life which will raise the average lumens produced per lamp but, since burnouts will not be replaced as they occur, the LBO factor will be lower. The tradeoffs should be considered in reaching a decision on the factors to be used in calculations.

Luminaire dirt depreciation (LDD) is the result of dirt and other airborne contaminants collecting on the luminaire reflective and enclosing surfaces.

IMPROVEMENT WITH MAINTENANCE

Maximizing each of the listed maintenance factors will improve lighting energy efficiency and cost by:

1. Reducing the initial number of installed fixtures.
2. Reducing the connected power load.
3. Reducing the energy consumed since one of the energy factors, power, has been reduced.
4. Reducing lamp replacement cost with fewer lamps to replace.
5. Reducing luminaire maintenance cost with fewer luminaires to maintain.

Maintenance will be easier with a good design, particularly in industrial areas where there is a reluctance to perform maintenance if it will affect production. Therefore, luminaire location, planned maintenance schedules for lighting and maintenance aids, such as luminaire-lowering devices, can be very effective in reducing maintenance cycles, maintaining maximum performance of the lighting system and, thereby, contributing to the energy efficiency of the lighting system.

CONTROL OF OPERATIONS

Most of the factors in lighting energy management covered here have been related to the "power" portion of the "energy" equation. The "time" factor relates to the capability to control the lighting system and the hours of operations. Improved energy management may be facilitated if the following measures are considered and implemented where practical:

1. Switch small groups of luminaires to permit operating only those which are necessary for the work being performed.
2. Provide multi-level control of the lighting system within an area by using alternate lamp and/or luminaire switching or multi-level ballasts.

3.   Group tasks to place those requiring higher illuminance together and then make use of non-uniform lighting to conserve energy.

4.   Use dimming systems to allow variable control of the lighting system for varying tasks.

5.   Integrate photocell control devices with switching and / or dimming controls to provide automatic control during periods of adequate daylighting.

6.   Install timers to automatically switch lights off during periods when they are not required.

7.   Implement power system kilowatt demand controllers to switch off (or down) noncritical lighting loads when the demand sensor approaches the present load limit.

8.   Restrict parking to selected areas on second and / or third shifts to reduce the parking lot lighting which must be operated during these hours.

9.   Arrange cleaning and maintenance to permit more effective use of lighting energy.

Much has been done in recent years to automate many of these energy conservation techniques. Programmable controllers, demand limiters and automatic timers are commonplace in many large facilities. These systems can be cost effective, particularly when the lighting control forms only a portion of the building systems controlled by these devices. There is a need, however, to perform in-depth cost studies to assure the effectiveness of the substantial initial investment required with some state-of-the-art energy management systems.

A decision cannot be made on the basis of energy consumed by lighting alone—the building heating and cooling costs, plus the productivity of the workers, must be factored into any evaluation. With every energy management system, the return on investment will be improved if the employer and employees are committed to making the system work effectively. Involving the employees in the purposes and operation of the energy management measures will give them a sense of contributing, not only to reducing the company's operating costs but to improving their own quality of life and conserving our vital natural resources.

References
1.   IES Lighting Handbook, 1981 Reference Volume, The illuminating Engineering Society of North America, New York, NY. 10017.

Chapter 3

Practical Selection of Fluorescent Lamps with Emphasis on Efficiency and Color

R.E. Snider

During the last few years numerous new fluorescent lamps have been introduced. Today, there is a wide selection of lamps of different wattages and colors available. The efficiency of these lamps can vary significantly. The purpose of this chapter is to investigate the choices which are available and give guidance in the practical selection of fluorescent lamps. Since most of the fluorescent lamps used today are 4' in length versus 8', this chapter will deal mainly with 4' lamps.

The concepts presented in this chapter are that of the author. Generalizations will often be made to present ideas in a practical manner. The author uses utility rates and construction data for Alabama. Each analysis should use the cost data for the specific location in question.

Although 4' lamps are available in other wattages we will be comparing 40W, 34W, and 32W lamps as these are the most popular. We will first compare cool white lamps since they comprise the majority of lamps sold. When comparing lumen output of lamps of the same wattage from three different manufacturers, they generally differed by 0-7%. Most lumen outputs differed no more than 3%. Since design is usually done with worst case situations, the minimum lumen output rather than the average lumen output was selected for comparison purposes. Prices for lamps, fixtures, labor rates, energy costs, etc. are approximate prices and are given only to allow comparisons to be made in a realistic manner. Prices, wage rates and labor costs may vary drastically from region to region and these variables should be adjusted for each geographic area.

In the economic analysis we will generally consider the number of fixtures, the installed cost of the fixture with lamps, replacing lamps every 5 years, and energy costs. Economic decisions will be based on the present worth method of comparison.

We will first compare 40W lamps versus 34W lamps. Comparisons will be made using economy and premium grade troffers with energy saving ballasts. Since the 40W lamp produces 15% more light than the 34W lamp, it will take 15% more 34W lamps to produce the same amount of light; thus 15% more fixtures. We will assume a cost of 4¢/kW hour. We will choose 3-lamp troffers since this is a good average between 2, 3, and 4 lamp troffers. We are concerned with the final cost to the owner and therefore will be using cost after the electrical contractor and general contractor have added overhead and profit. We will consider a combined mark-up of 30%. We will assume it takes 1 manhour to install one fixture at a labor rate of $9.50/hour. The average life of this type of lamp is 20,000 hours. Assuming 12 hours/day and 6 day/week operation, the lamp life is 5.3 years. We will assume a 5-year life. An interest rate of 8% compounded annually will be used.

	Economy Grade Troffer		Premium Grade Troffer	
	34W	**40W**	**34W**	**40W**
Initial Cost	+20%	—	+19%	—
5-Year Cost	+7%	—	+8%	—
10-Year Cost	+5%	—	+6%	—
15- Year Cost	+5%	—	+6%	—

Therefore, based on the assumptions we have made, 34W lamps are not economical for this area. Where energy costs are considerably higher they will be more economical.

Next we will compare the 40W/CW lamp to the standard size 32W/T-12/CW lamp. For simplicity only premium grade troffers will be used. If premium grade fixtures prove economical, then economy grade fixtures will also be economical.

It should be noted that 32W/T-12 lamps operate a little differently from standard 34W and 40W lamps. If the lamps are switched on from a cold start and remain on, the operation is the same. However, if the lamps are switched off and then switched back on within one minute, it may take 1-5 minutes for the lamps to restart and come back to full brightness. This could be critical if the lamps are on an emergency circuit connected to a generator and expected to be back on in no less than 10 seconds. It also causes nuisance problems by people who are not familiar with their operation. For example, if they turn the lights off, remember they have forgotten something and turn the light switch back on, the lamps will be very dim and have uneven light output. They may think something is wrong and call maintenance.

	40W	32W
Initial Cost	—	+24%
5-Year Cost	—	+9%
10-Year Cost	—	+7%

Here again 40W lamps appear the most economical for this area.

Next we will compare the standard 40W/CW to the 32W/CW/T-8. The T-8 lamp is similar to the standard 40W/T-12 except the T-8 is 1" in diameter rather than 1/2." Since there is less lamp to block the reflected light, the fixture is therefore more efficient. Coefficients of utilization will be higher for T-8 lamps than T-12 lamps. The fixture is about 9% more efficient with T-8 lamps versus T-12 lamps. Because of this a 40W/CW can virtually replace a 32W/T-8 lamp. The T-8 lamps require a different ballast than the standard lamps. T-8 lamps will not work properly on ballasts for standard lamps and vice versa. Also 4-foot T-8 lamps are only available in 32W. The 40W/T-8 lamps are S feet in length. You can expect an adder to the fixture cost where T-8 lamp ballasts are required. We will first consider magnetic ballasts and assume an adder of $7.50 per ballast (contractor cost) over energy saving ballasts.

	40W	32W/T-8
Initial Cost	—	+30%
5-Year Cost	—	+5%
10-Year Cost	—	+0%
15-Year Cost	—	+1%

Again the standard 40W lamp appears the most economical choice for this area. The increase in cost between 10-15 years is due to the fact that the ballast needs to be replaced at the end of 10 years.

Next we will consider electronic ballasts to see if this effects lamp choice. Some electronic ballast manufacturers make ballasts which will operate 1 or 2 lamps while others make an electronic ballast which will operate 1, 2, 3, or 4 lamps.

We will first compare a fixture with standard 40W/CW lamps and 2 energy saving ballasts versus a fixture with standard 40W/CW lamps and 2 electronic ballasts. We will assume an adder of $22 per ballast (contractor cost) over energy saving ballasts.

	40W/ESB	40W/Electronic(2)
Initial Cost	—	+66%
5-Year Cost	—	+22%
10-Year Cost	—	+11%
15-Year Cost	—	+1%

40W lamps with energy saving ballasts were again more economical.

We will not compare a fixture with standard 40W/CW lamps and 2 energy saving ballasts versus a fixture with standard 40W/CW lamps and one 3-lamp electronic ballast. We will assume an adder of $24 (contractor cost) for the electronic ballast over the fixture with energy saving ballast.

	40W/ESB	40W/Electronic(1)
Initial Cost	—	+36%
5-Year Cost	—	+4%
10-Year Cost	+5%	—
15-Year Cost	+14%	—

If you can afford the 36% increase in initial cost, the single electronic ballast should start saving money in 7-8 years.

34-watt lamps are not recommended for use on some electronic ballasts because of starting problems and therefore will not be considered.

Next we will compare the standard 40W/CW lamp versus the 32W/CW/T-8 lamp with 1 electronic ballast. The single electronic ballast costs only a little more for the T-8 lamps versus the T-12 lamps. We will assume an adder of $26 (contractor cost) over energy saving ballast.

	40W	32W/T-8/Electronic(1)
Initial Cost	—	+47%
5-Year Cost	—	+6%
10-Year Cost	+3%	—
15-Year Cost	+12%	—

Again the initial cost of the electronic ballast is extremely high, but after 8-9 years it should start paying off. Also note that the standard 40W lamps with electronic ballasts payoff sooner with a lower first cost than the 32W/T-8 lamps.

Note: A lamp life of 20,000 hours using electronic ballasts has been assumed. Lamp life may need to be reduced to 15,000 hours depending on the lamp and ballast combination used.

Up to now we have assumed energy saving ballasts to be more cost effective than standard ballasts. Let us verify this assumption. We will assume a top-of-the-line energy saving ballast with an adder per ballast of $2.50.

	Economy Grade Troffer		Premium Grade Troffer	
	40W/Standard	40W/E.S.B.	40W/Standard	40W/E.S.B.
Initial Cost	—	+12%	—	+8%
5-Year Cost	+3%	—	+3%	—
10-Year Cost	+6%	—	+5%	—

Therefore, the energy saving ballast costs more initially but pays off in less than 5 years and is a good choice.

Generally we have found that energy saving products are not always the most economical. This is because more fixtures will usually be required with the energy saving lamps. The cost of the lamps is also usually higher. When using 32W/T-12 lamps remember they have different warm start characteristics than standard lamps. 32W/T-8 lamps also require special ballasts and are not compatible with standard ballasts. Some fixtures will have higher coefficients of utilizations when using T-8 lamps versus T-12 lamps. 34W lamps are not recommended to be used with some electronic ballasts and these lamps may have lower lamp life.

SUMMARY OF FIXTURES WITH DIFFERENT TYPE CW LAMPS

	40W	34W	32W	32W/T-8	40W/ Elect.(2)	40W/ Elect.(1)	32W/ T-8 Elect.(1)
Initial Cost	—	+19%	+24%	+30%	+66%	+36%	+47%
5-Year Cost	—	+8%	+9%	+5%	+22%	+4%	+6%
10-Year Cost	—	+6%	+7%	+0%	+11%	−5%	−3%
15-Year Cost	—	+6%	−2%	+1%	+1%	−14%	−12%
Watts Used	136	115	109	108	107	97	88
No. of Fixtures	1.0	1.15	1.17	1.0	1.0	1.0	1.0

NOTE: *The above data are based on utility rates and construction costs for Alabama. Each job will require a cost analysis based on the specific application and locality. Figures may differ significantly from those shown above.*

COLOR CONSIDERATIONS

The color of fluorescent lamps is generally described by the correlated color temperature expressed in Kelvin degrees and the color rendering index of C.R.I. expressed as a whole number from 1-100.

The International Commission on Illumination (CIE) developed a diagram called the CIE Chromaticity Diagram to aid in specifying colors. Plotted on the diagram is a curved line called the Black Body Locus. This line represents the change in color of a piece of metal as it is heated. Therefore, a particular color can be specified by indicating the degree Kelvin provided it is an incandescent lamp.

The spectrum from fluorescent lamps does not fall exactly on the curve; therefore, the correlated color temperature must also be specified. The problem with this method is that one lamp could plot above the curve where another plots below the curve. Both could have the same correlated color temperature while having different color characteristics. Therefore, temperature is not an exact measure of color for fluorescent lamps.

To aid in specifying colors, a system called the color rendering index (CRI) was developed. The purpose of the color rendering index is to measure the shift from the curve on the CIE chromaticity diagram. This is done by measuring the shift of eight specified colors compared to a reference source of the same degree Kelvin. The shift in each color is determined and then the numbers are averaged to determine the CRI. Therefore, one lamp could score high in reds and another high in blues and yet have the same average score. This is not common but it can happen.

Though cumbersome, sometimes the best method to specify fluorescent lamps is to examine the color spectrum of a lamp by its spectral power distribution curves. From these curves it is easy to identify that cool white lamps have more blue and less red than warm white lamps. Also, you can see that deluxe cool white lamps are much stronger in reds than cool white or warm white. These lamps have a more even distribution of colors but are not as good as the 5000°K lamps.

From the spectral distribution curves you can see the difference in the single coated lamps and the fairly new double coated tri-phosphor lamps. The double coated lamps have one coat of the same phosphors as the regular lamps and then they have a second coat of more expensive phosphors. The second coat of phosphors contains three strong primary colors producing phosphors in the yellow-red, green, and blue-green wave lengths. The manufacturer's design philosophy is that most colors can be made by a combination of these three colors and therefore the color rendering ability of the lamp is improved. These lamps provide an economical way of improving the CRI while maintaining high lumen output. These are available at 3000°K, 3500°K, and 4100°K.

In choosing a fluorescent lamp there is no right or wrong color. We would like a fluorescent lamp which renders colors as we are used to viewing them. For example" at home we are normally viewing objects from a 3000°K source; at work from a 4000°K source and outside the daylight varies from 1800°K to 28,000°K.

A lot of objects simply do not appear as pleasing when viewed under natural daylight as they do under other light sources. Therefore selection of lamps is a matter of personal preference in most cases.

In the following section, 10 lamp color comparisons will be made. Comparisons will include temperature, CRI, lumens, lamp cost and 5-year cost for the following:

1. CW vs WW	6. CWX vs 5000°K
2. CW vs LW	7. CW vs SP35
3. CW vs 5000°K	8. WW vs. SP35
4. WW vs LW	9. CW vs SP41
5. WW vs 5000°K	10. WW vs SP30

No. 1: CW vs WW

	Watts	Approx. Temp.	Approx. CRI	Approx. Lumens	Approx. User Lamp Cost	No. of Fixtures	Approx. 5-Year Cost
CW	40	4200	62	3150	2.50	+2%	+1%
WW	40	3000	52	3200	3.30	—	—

Remarks: Clothes and skin will generally look better under WW. WW gives skin a slight tan and brings out the reds in clothes.

No. 2: CW vs LW

	Watts	Approx. Temp.	Approx. CRI	Approx. Lumens	Approx. Lamp Cost	No. of Fixtures	Approx. 5-Year Cost
CW	404200	62	3150	2.50	—	—	
LW	344200	49	2925	3.75	+8%	+2%	

Remarks: CW definitely has better color. LW is not available in 40W. The LW has a high lumen per watt output but still does not prove economical according to the figures used.

No. 3: CW vs 5000°K

	Approx. Watts	Approx. Temp.	Approx. CRI	Approx. Lamp Lumens	No. of Cost	Approx. 5-Year Fixtures	Cost
CW	40	4200	62	3150	2.50	—	
5000°K	40	5000	90	2000	6.20	+43%	+52%

Remarks: The 5000°K has excellent color rendering characteristics throughout the spectrum. The 5000°K lamp is generally used where color determination is critical.

No. 4: WW vs LW

	Watts	Approx. Temp.	Approx. CRI	Approx. Lumens	Approx. Lamp Cost	No. of Fixtures	Approx. 5-Year Cost
WW	40	3000	52	3200	3.30	—	—
LW	34	4200	49	2925	3.75	+9%	+10%

Remarks: Although the CRIs are close, there is a big difference in the lamps' rendering. The LW appears bland while the WW is rich in red.

No. 5: WW vs 5000°K

	Approx. Watts	Approx. Temp.	Approx. CRI	Approx. Lamp Lumens	No. of Cost	Approx. 5-Year Fixtures	Cost
WW	40	3000	52	3200	3.30	—	—
5000°K	40	5000	90	2200	6.20	+45%	+52%

Remarks: The WW may be considered more pleasing while the 5000°K is considered more natural.

No. 6: CWX vs 5000°K

	Watts	Approx. Temp.	Approx. CRI	Approx. Lumens	Approx. Lamp Cost	No. of Fixtures	Approx. 5-Year Cost
CWX	40	4100	89	2100	4.50	+5%	+2%
5000°K	40	5000	90	2200	6.20	—	—

Remarks: Both have excellent color rendering characteristics, and yield fairly "natural" colors. They do not over-emphasize anyone color. The 5000°K has a more even color balance and should be used where color is critical such as in nurseries in hospitals. CWX should be good enough for most exam rooms and treatment rooms in hospitals.

No. 7: CW vs SP35

	Watts	Approx. Temp.	Approx. CRI	Approx. Lumens	Approx. Lamp Cost	No. of Fixtures	Approx. 5-Year Cost
CW	40	4200	62	3150	2.50	+1%	—
SP35	40	3500	69	3180	4.10	—	+2%

Remarks: The SP35 color rendering is usually more pleasing than the CW. The SP35 offers a good compromise between the red color of WW lamps and the blue color of CW lamps. The SP35 does a good job on clothes and skin. The popularity of this lamp is just beginning to catch on and as demand increases the price will probably drop. The SP35 may be the lamp of choice for general use.

No. 8: WW vs SP35

	Watts	Approx. Temp.	Approx. CRI	Approx. Lumens	Approx. Lamp Cost	No. of Fixtures	Approx. 5-Year Cost
WW	40	3000	52	3200	3.30	—	—
SP35	40	3500	69	3180	4.10	+1%	+2%

Remarks: Again the SP35 is a good compromise between the WW and CW.

No. 9: CW vs SP41

	Watts	Approx. Temp.	Approx. CRI	Approx. Lumens	Approx. Lamp Cost	No. of Fixtures	Approx. 5-Year Cost
CW	40	4200	62	3150	2.50	+3%	+1%
SP41	40	4100	69	3240	3.75	—	—

Remarks: The SP41 offers some improvement in color over the CW. This may be a good lamp if you are concerned that the lamps will be mixed with CW lamps when replacement starts. The lamps will look the same in the fixture. This may also be a good lamp for patient bedrooms. Each bedroom has a window and this compromise between incandescent and daylight is okay.

No. 10: WW vs SP30

	Watts	Approx. Temp.	Approx. CRI	Approx. Approx. Lumens	Lamp Cost	Approx. No. of Fixtures	5-Year Cost
WW	40	3000	52	3200	3.30	+1%	
SP30	40	3000	69	3230	4.10	—	

Remarks: The SP30 has a significantly improved CRI and is an excellent choice if you want a 3000°K lamp. Great for clothes and skin color enhancement.

The type of a lamp should not be chosen to highlight the color of wallpaper or carpet in a room. More important is the function of the space and what will be in the room. The designer must consider the overall effect of the space when selecting a lamp. There is no right or wrong choice. It is simply a matter of opinion. The author feels that the SP35 lamps are probably the best choice for general offices and variety retail stores and the SP30 lamps for clothing stores.

The prices used in the calculations for this chapter are based on the approximate cost of materials, labor, and energy in Alabama. These factors will vary significantly from area to area and must be adjusted accordingly. This chapter has illustrated the various options available when selecting a fluorescent lamp.

Chapter 4

Ideal Task Lighting Arrives!

Donald R. Wulfinghoff, P.E.

The lighting technology introduced here is covered by U.S. Patent 7,824,068, "Lighting Fixtures and Systems with High Energy Efficiency and Visual Quality."

A REVOLUTION IN ENERGY EFFICIENCY AND VISUAL QUALITY

Lighting is one of the three largest energy uses in buildings, and often the largest. We have seen the satellite images of the earth at night, which reveal how much of humanity's energy resources are expended for lighting.

Lighting that is used at localized activities, such as desks, reading chairs, bench work, and machine tools, comprises a large fraction of this consumption. This category of lighting offers enormous potential for improving efficiency. It is easy to calculate that contemporary lighting for localized activities typically consumes about one hundred times more energy than is theoretically needed for effective and comfortable vision.

Today, the largest part of lighting energy waste is geometrical. It occurs from wasting light in places where it is not needed and from distributing light with insufficient regard to effective vision. The first cause has long been recognized, but not the second.

The conceptual solution to geometrical energy waste in lighting is "task lighting." At its most basic, task lighting concentrates light within the areas requiring illumination. This concept was popularized during the 1970's "energy crisis" years. However, the task lighting of that era was not accepted by users, because it did not satisfy the requirements for good visual quality. Figures 1 and 2 illustrate its problems.

As a result, interest in task lighting withered, and it has remained a poorly defined concept. ASHRAE/IESNA Standard 90.1[1], which is the de facto energy conservation standard for the United States, still cites task lighting only as "supplemental," and it referred only to "undershelf or undercabinet" configurations. Its great potential for improving the energy efficiency of buildings remains untapped.

Task lighting stalled because it was misled by wrong assumptions. One was that task lighting fixtures should be installed within or near the activity. Another was using small fixtures of narrow aperture. As a result, task lighting became associated with virtually all the ills that can afflict lighting, including glare, veiling reflections, non-uniform illumination, and harsh shadows.

IDEAL TASK LIGHTING IS ACHIEVED

But finally, fully satisfactory task lighting has been achieved. After decades of analysis and experimentation, we have found that there is one way to arrange task lighting that combines high energy efficiency and per-

Figure 1. The bad old days of task lighting, as it was first promoted for energy conservation. A desk lamp is located within or near the activity area, where it creates extreme contrast and limited range of illumination.

fect visual quality. We give this configuration a name, "ideal task lighting." We define it as illumination of an activity area in a way that optimizes all requirements for visual effectiveness and comfort[2]—without compromise—and that minimizes energy consumption.

And here it is! Figure 3 shows how well ideal task lighting can perform in a challenging application. The activity area in the photograph includes an L-desk, a long work table, a printer that is beyond the desk, a shelf unit on the work table, and a tall card file on the shelf unit. All are perfectly illuminated.

Figure 4 shows ideal task lighting from the user's perspective. Note the ample, uniform illumination of every object within the activity area. Also, note the complete absence of glare, veiling reflections, and harsh shadows.

Figures 3 and 4 illustrate two major breakthroughs. First, with respect to visual quality, *the user enjoys perfect illumination throughout the activity area.*

Second, with respect to energy efficiency, *the total power to the lamps in this application is only 29 watts.* This is approximately one third of the power

Figure 2. Old fashioned task lighting, from the user's perspective. It aggravates all major interferences with vision, including glare, veiling reflections, extremely non-uniform illumination, and harsh shadows. The resulting visual discomfort caused users to abandon task lighting.

prescribed by ASHRAE/IESNA Standard 90.1[4], if the lighting operates continuously.

In addition, ideal task lighting allows the lighting for individual activity areas to be turned off without hindrance to the lighting of adjacent activity areas. Thus, in an environment with multiple activity areas, the improvement in energy efficiency can be several times greater than a factor of three.

The figures show how little light escapes from the activity area. This is the essence of task lighting. Other areas within the room have their own

Figure 3. Ideal task lighting. Compare this to Figure 1. In this exceptionally large activity area, the level of illumination for each part of the scene can be selected by the user. There are 11 LED lamps in each fixture, with a total lamp power of only 29 watts for the entire scene. Translucent glare shields provide some glow on the ceiling and upper walls, and the fixtures may be adapted to provide other decorative effects.

lighting, tailored to the requirements of each area. (For the photographs, all other room light was turned off.)

Ideal task lighting has been tested for over three years, using a variety of equipment, in three especially relevant applications: a large office work station, a small desk, and a reading chair. All three applications demonstrated ideal visual quality with dramatically reduced energy input.

WHERE IDEAL TASK LIGHTING SHOULD BE USED

Ideal task lighting should be the preferred method of illumination for all applications where it can fulfill its potential for visual quality and energy efficiency. In general, these applications are activities that occur within a fixed area, where the viewer is situated in a specific location with respect to the activity.

Probably the largest application category is work at desks. In office

Figure 4. Ideal task lighting from the user's perspective. Compare this to Figure 2. All sources of visual discomfort have been eliminated. The scene has no glare, no veiling reflections, and no objectionable shadowing. The illumination can be uniform, or the user can highlight or dim parts of the scene as desired.

buildings, ideal task lighting logically should become the primary form of lighting.

Another large application category is reading while seated, in both commercial and residential settings. Other application categories include workbenches, bank teller stations, retailing that uses localized vending areas, and some food preparation activities. Typical industrial applications are work at machine tools and stations on a production line. Residential applications extend to hobbies, sewing, and workshop activities.

Ideal task lighting is tailored to each activity area, and each activity area is controlled independently. A single person may use more than one activity area. And, one activity area may be shared by different persons at different times.

For example, in a large office with many desks and shared file cabinets, each desk has its own task lighting and the file cabinets have their own task lighting. Each activity area might have a people sensor, so that light is never wasted on inactive space.

WHAT'S NEW ABOUT IT?

The undesirable aspects of old fashioned task lighting stem from bad geometry and from the inflexibility of the equipment. Ideal task lighting overcomes these problems with the following features, which are illustrated in Figure 5:

- The fixtures are located outside the activity area, situated so that they cannot create uncomfortable visual conditions. By the same token, ideal task lighting can be mounted outside the activity area (e.g., from the ceiling), avoiding interference with the activities or layout of the lighted area.

- The fixtures have a large aperture, or average width, in relation to the distance of the fixture(s) from the activity area. This allows each fixture to illuminate the activity from many directions. The fixtures achieve a large aperture by using multiple small light sources (as in Figures 3 and 5); or, by using a single light source in combination with a reflector and / or lens(es) to direct light from multiple directions; or, by a combination of these.

- The fixtures allow the user to tailor the illumination to the activity. In fixtures that use multiple lamps, the individual lamps may be separately adjustable in direction, and perhaps, in intensity. For example, each lamp of the fixtures in Figures 3 and 5 can be aimed in any direction. In fixtures that use reflectors or lenses for directing the

Figure 5. Key features of this ideal task lighting configuration:
- The fixtures are located outside the activity area.
- The fixtures are located above, to the side, and somewhat behind the primary orientation of the user.
- Each fixture has an aperture that is large in relation to the distance from the fixture to the illuminated surface.
- The fixtures efficiently contain light to the activity area.
- The fixtures allow tailoring of the illumination level throughout the scene.
- The fixtures use lamps that are efficient at low light output.
- For wide activity areas, such as this desk, the fixtures are used in pairs.

This arrangement is the same as in Figure 3, except that the glare shields are almost opaque, which can make the fixtures almost invisible, if desired. Again, the total lamp power for this large activity area is only 29 watts.

light from individual lamps, the illumination pattern can be adjusted by moving the reflectors or lenses with respect to the lamps, and in other ways.

HOW IDEAL TASK LIGHTING ACHIEVES
EXTREME ENERGY EFFICIENCY

The extremely high energy efficiency of ideal task lighting is achieved by making all these improvements[3]:

- optimized light distribution. The new technology can limit lighting to the activity area very precisely. Further, the light distribution within the activity area can be tailored to the levels desired by the user, avoiding energy waste within the activity area.

- minimal light loss between the light sources and the illuminated activity. Light travels directly from the light sources to the illuminated activities. This eliminates light absorption within the lighting equipment itself and by the surroundings, which causes major energy waste in most lighting. Some variations of the technology use reflectors or lenses, which have low light loss.

- reduced illumination level requirements. By eliminating all factors that interfere with human vision, the technology makes it possible to significantly reduce the illumination levels that are needed for effective vision. (This important physiological effect was discovered during the development of the invention.)

- efficient utilization of LED's and other light sources that are limited to low power input. Ideal task lighting utilizes LED's to their maximum potential, turning their individually low power into a virtue. For example, each of the fixtures in Figures 3 and 5 uses eleven LED's.

- making each occupant's lighting independent in space and time. This was the original defining characteristic of task lighting. It provides an enormous efficiency improvement over area lighting by allowing lighting energy to be reduced in proportion to the occupancy of the space or to the use of an activity. By bringing the concept of task lighting to fruition, ideal task lighting finally makes it possible to exploit this benefit.

HOW IDEAL TASK LIGHTING ACHIEVES
PERFECT VISUAL QUALITY

Visual quality is optimized by eliminating all the factors that cause visual discomfort and interference with effective vision. Specifically:

- glare is completely eliminated. The fixtures are located outside the part of the user's visual field that is susceptible to glare.

- veiling reflections are completely eliminated. The fixtures are located with respect to the activity and the viewer so that veiling reflections are impossible.

- undesirable variations in illumination within the activity area are eliminated. The light distribution is tailored to the illumination requirements of the activity and to the preferences of the viewer.

- excessive self-shadowing is eliminated by the crossfire of illumination from multiple directions.

- harshness is eliminated by the large effective aperture of the fixtures.

HOW IDEAL TASK LIGHTING CAN *ENHANCE DECOR*

Ideal task lighting can be an attractive decor feature in ways that are limited only by the imagination of the manufacturer, the interior designer, and the lighting designer. Examples are:

- arranging LED light sources around the perimeter of the light fixtures in a decorative pattern, with a decorative holder

- using LEDs, tinted glare shields, or other light sources to create patterns of color

- If the interior designer wishes to subdue the appearance of the fixtures, using an appropriate frame color and opaque glare shields can make the fixtures blend into the ceiling theme.

For example, the glare shields in Figure 3 are translucent, so they can be used as a major element of the decor. In contrast, the fixtures in Figure 5 use glare shields that are almost opaque, so they blend into the background.

A NEW APPROACH TO LIGHTING DESIGN

Exploiting ideal task lighting requires a change in lighting design philosophy. In contemporary lighting, the design is completed before the building is built, based on a doctrine of area lighting. This leaves building owners and occupants unable to adapt lighting to individual needs, so they live with high energy cost and with illumination that is less than optimum.

Because contemporary lighting design doctrine disregards the needs of individual users, occupants must arrange portable fixtures to satisfy their individual requirements. Or, vast amounts of energy are wasted to illuminate vacant space to the same standards as activity areas.

But now, with ideal task lighting, lighting design becomes a collaboration between the designer, the manufacturer, and the user. The manufacturer produces fixtures that have the desired combination of aperture, lamp types, adjustability, mounting options, and decorative features. The designer specifies the fixtures, the mounting method, and the power system. The locations of the activity areas determine where the fixtures will be installed, usually as the building is being occupied. Installers and/or users (depending on the mounting method) locate the fixtures properly with respect to each activity area. And finally, the users are able to adjust their fixtures to tailor the illumination precisely to their preferences.

FOLLOW THE INSTALLATION MANUAL

The revolutionary advantages of ideal task lighting are primarily the result of optimum lighting geometry. Achieving its benefits requires placement of the fixtures as illustrated in Figures 3 and 5. These are the installation instructions.

Two Fixtures, on Opposite Sides of the Activity Area

For most applications, ideal task lighting requires two fixtures for each activity area. This follows from a chain of logic. To avoid veiling reflections, the light must come from the side. But, if light comes from the side, a single fixture cannot avoid extremely unbalanced light distribution and excessive self-shadowing. A symmetrically placed pair of fixtures, in combination with the features that follow, can avoid all three problems simultaneously. The overlapping distribution patterns of the two fixtures can be adjusted to provide uniformity of illumination, along with any desired variations in illumination level. The opposing geometry of the two fixtures also eliminates self-shadowing.

The separation between the fixtures is determined by the width of the activity area, or by the body of the viewer, whichever is wider. For desk and shop work, the fixtures are located over the ends of the activity area.

One significant exception is lighting for reading chairs, where a single fixture may suffice. This is because the illuminated area (typically a book or magazine) is narrow and because the viewer's hands are not located above reading material.

Location above the Viewer's Head

The bottom of each fixture should be located at least one foot above the head of the user. For typical applications, mounting below the ceiling is ideal. The only limitation on height is that the aperture of the fixture should be about 25% to 35% of the distance from the fixture to the surface of the activity area.

The overhead location eliminates light source glare for the user, it contributes to the muting of self-shadowing, and it provides uniform illumination of desks and other horizontal activity areas. It exploits the shape of the human skull, including eyebrows and limited lateral vision, to avoid glare. The fixture height also allows the viewer to turn his head for work at the sides of the activity area.

Most work surfaces are horizontal, or nearly so. By shining light on horizontal surfaces from a predominantly overhead direction, the overhead location helps to provide uniform light distribution on such surfaces. (Computer screens are nearly vertical, but they do not require illumination.)

Location Slightly Behind the Viewer

Over-the-shoulder location eliminates veiling reflections when looking toward the sides of the activity area. It helps to eliminate light source glare. And, it expands the illumination of side areas, such as the extensions of L-desks.

The fixtures can be moved more rearward to accommodate activities to the sides of the viewer. If this is done, the fixtures should be mounted higher to compensate. The limitation is that the body of the user will eventually block the crossfire that mutes self-shadowing. Also, if the fixtures are moved too far back, reflections of the fixtures may become visible in a computer screen, but this can be avoided by tilting the screen to a more vertical orientation.

Glare Protection for Outside Viewers

It is vital to keep the fixtures for one activity area from creating glare at a different activity area, or for bystanders. One effective method of blocking bystander glare is to equip the fixtures with integral glare shields, as in Figures 3 and 5. These eliminate glare to the side and rear of the activity area. Fixture-mounted glare shields may not suffice for bystanders who face the activity area from the front.

Another reliable technique is to install a separate glare shield that is located between the activity area and potentially vulnerable bystanders. Typically, this is easily mounted from the ceiling. For example, the yacht ensign in Figure 3 serves this purpose, except that it would be wider.

The same kind of glare shield also protects the user of the activity area from any sources of glare outside the activity area, such as a window that faces the sun. For example, the yacht ensign was originally installed to block the glare of a conventional ceiling fixture located ahead of the desk.

UNLIMITED ADAPTATIONS

The new technology of ideal task lighting introduces a set of geometrical principles that cannot be altered. However, many variations of this patented technology[5] are possible without violating those principles, including options for:

- mounting the fixtures,
- power sources,
- types and numbers of lamp units,
- lamp holder design,
- the arrangement of light sources on lamp holders and the arrangement of lamp holders on fixtures,
- methods of aiming and focusing the light,
- the aperture of lamp units,

- the beam width of lamp units, and
- shielding of external viewers against glare.

As we observed previously, ideal task lighting opens up a host of opportunities for exciting new decorative concepts.

SO, HOW WELL DO USERS LIKE IT?

Shortly after the test program commenced, the test subjects noted that the experimental lighting was notably more appealing than the original conventional lighting.

The test program also revealed that ideal task lighting initially looks and feels different from contemporary lighting. This period of novelty seems to last from several hours to several days. We can imagine that similar adaptive responses occurred during the change from whale oil lamps to gas lighting, and from gas lighting to electric lighting.

The illumination provided by ideal task lighting is different from the lighting that people usually encounter indoors, from either daylighting or contemporary artificial lighting. The complete absence of veiling reflections, glare, and harsh shadows is rarely encountered in daily experience. Initially, the task area lighting may seem unusually clear and uniform (but not flat).

An important finding is that viewers are comfortable with significantly lower measured illuminance levels with ideal task lighting than with conventional lighting. (For test subjects in the mid-to late 60's age range, 250 lux typically was reported as comfortable for desk work and reading.) We attribute this to the absence of veiling reflections and glare, and to the uniformity of illumination. I.e., no useful light is wasted to overpower light that interferes with vision. A related effect is that the activity area lighting may appear dim to an outside observer.

There's one final issue. With ideal task lighting, it is possible to keep less than 5% of the light from falling outside the activity area. No discomfort or annoyance results from this. Part of the dogma of contemporary lighting design is that people become uncomfortable if they are located inside an "island of light" that is surrounded by darkness. But, in decades of working with lighting, we have never observed this alleged phenomenon.

References

1. ANSI/ASHRAE/IESNA Standard 90.1-2004, paragraph 9.4.1.4 (d).
2. Donald R. Wulfinghoff, Energy Efficiency Manual, (Energy Institute Press, 2000), Reference Note 51, "Factors in Lighting Quality," page 1427.
3. Wulfinghoff, op. cit., Section 9 Introduction, page 1019.
4. ANSI/ASHRAE/IESNA Standard 90.1-2004, Table 9.6.1.
5. U.S. patent 7,824,068, "Lighting Fixtures and Systems with High Energy Efficiency and Visual Quality," issued November 2, 2010.

Chapter 5

Efficient and Effective Lighting Design

Paul L. Villeneuve, P.E.

OVERVIEW

Lighting design traditionally focused on providing adequate lighting to support activities. The design phase mainly consisted of using estimates of luminaire output to locate the luminaire installation location. Computer programs have been developed to provide much finer lighting layout. This chapter attempts to give the reader an introduction to the process of lighting design.

Keywords: Lighting Design, Luminaire Specification, Fixture Configuration.

INTRODUCTION

Artificial lighting permits humans to increase efficiency, comfort, and safety. A well designed and implemented lighting project not only meets criteria for the particular task or activity to be performed but does so efficiently while minimizing the impact on the environment. In addition, artificial lighting can provide a greater overall sense of wellbeing particularly when natural lighting is limited.

Lighting layout and design has evolved primarily due to the development of lighting design software. Traditionally, lighting designers would locate luminaires based on the designers experience and the anticipated light output. This method resulted in widely varying lighting levels and higher energy consumption. Lighting design software provides greater flexibility for lighting designers by performing light level calculations in a simulated environment. The designer can tailor the design to maximize light output and efficiency.

KEY ELEMENTS OF LIGHTING DESIGN

A lighting design focuses on the following elements:
* Extending the opportunity to perform activities
* Enabling a perception of better security
* Providing enhanced viewing for better accuracy
* Maximizing the conversion of energy to visible light
* Offering a more comfortable environment

Each of these elements must be weighed by the designer to determine the most effective, economical, and efficient design for a particular application

LIGHTING BASICS

Lighting designers utilize a few key elements when designing lighting systems. This section presents them with a brief description.

Illumination Levels

Illumination is measured using the foot-candle or lux. The definite of a foot-candle is the light level falling on a 1 ft^2 surface one foot away from a standard candle. General illumination levels are specified by design objectives. The Illuminating Engineering Society (IES) produces guidelines for suitable illumination levels dependent upon the environment and activity to be performed. In particular, the Lighting Handbook produced by IES is an excellent reference. The reader is encouraged to review this helpful document.

A frequently encountered problem is to over specify lighting resulting in excessive lighting and over illuminating an area can occur. Overly lit areas can create several health impacts including headaches, fatigue, and stress and anxiety. Previous lighting designs commonly over lit areas.

Luminaire Temperature

Light is electromagnetic energy that is vibrating a certain range of frequencies that are visible to the human eye. The visible light energy vibrates from about 400 to 790 THz. The color of the light varies as the frequency changes. For example, the color red is approximately

400 THz whereas the color violet is approximately 780 Thz.

Lighting designers do not discuss color using frequency or even wavelength. Rather, the concept is based on a blackbody radiator. A perfect blackbody radiator does not emit or absorb radiation at low temperatures and therefore does not have any color. As the black-body radiator is heated, it begins to emit thermal radiation and give off color. Lighting designers use the concept of a blackbody radiator to describe the color of a luminaire. Typical luminaire temperatures include 2700°K for high pressure sodium and 4000°K for metal halide. This is indicative of a yellowish light for the high pressure sodium and a white light for the metal halide.

Color Rendering Index

Another important factor for a luminaire is the ability of an observer to distinguish different colors of objects. To quantitatively express the ability of a luminaire to faithfully replicate color, the International Commission on Illumination developed a method to describe the color replication ability. This method is entitled the Color Rendering Index (CRI). The light emission from the sun is considered full spectrum and any color can be distinguished in natural sun light. As a result, it is indexed as a perfect score with a CRI of 100. In general, incandescent lights have a perfect score as well. Artificial light may not output the full spectrum of light. Monochromatic lights (lights that produce few wavelengths) have low CRIs. For example, high pressure sodium lights have a CRI of around 25. Metal Halide lights have a CRI of around 60. High CRIs are important for certain activities that require accurate description of color. For example, purchasing a vehicle in the evening requires the buyer to be able to ascertain the correct color.

Efficiency

The ability of a device to convert input energy into a desired output energy determines the efficiency of an object. From a lighting designer's perspective, achieving high efficiency was somewhat assured by using high intensity discharge type fixtures. Current design objectives are starting to focus on ensuring high efficiency while meeting illumination levels. For indoor luminaires, efficiency levels range from approximately 2.5% for standard incandescent luminaires to 14% for fluorescent fixtures. For outdoor luminaires, efficiencies range from 15% for metal halide fixtures to 20% for high pressure sodium. Although the high pressure sodium luminaire has the highest efficiency its CRI and color temperature are such that these luminaires have limited application. At the time of the writing of this chapter, LED luminaires do not have extensive use. However, efficiency ranges for LED luminaires are from approximately 9% to 15%.

In addition to designing a luminaire with an efficient ballast and lamp to convert input energy to visible light, designing the refractor, the luminaire globe, or lens, is an important aspect of luminaire design to maximize efficiency. The refractor should absorb or filter as little light as possible. Clear refractors are possible but they can create glare as the lamp is readily viewable. Special designed refractors appear opaque and allow most of the light through and limit glare.

Lighting distribution

One of the most important factors affecting luminaire selection is specifying the distribution of light from the luminaire. A typical light bulb for a particular luminaire produces light in all directions. To meet the desired light distribution, manufacturers spend a great deal of effort designing reflectors within the fixture. Indoor fixture reflectors are generally simpler than outdoor reflectors. Indoor reflectors primarily reflect indirect light into the workspace to minimize glare from the light bulbs. Outdoor fixtures are generally compact as compared to the output wattage and careful design of the reflector is required to ensure even distribution. Outdoor luminaires can be specified with a drop-type lens. In this case, a portion of the light bulb is below the base of the luminaire. This results in greater light distribution but also results in glare as the bulb is visible from multiple locations. For further details on standard and non-standard lighting distributions please refer to IES documents.

A concern regarding light distribution is light trespass. Light trespass results from light impacting adjoining properties. Light trespass is only applicable for outdoor luminaires. Many luminaires can be supplied with a blocking plate preventing a significant portion of the light from projecting in a specific direction. This type of plate is called a house side shield. Use of a house side shield should generally only be used at owner property lines. Light trespass can also occur if light from a fixture bleeds into the sky. Light trespass into the sky has various impacts that are difficult to quantify. The International Dark-Sky

Association provides certifications for fixtures that meet design criteria

Another factor affecting lighting distribution is the mounting height for which the luminaire(s) is/are mounted. The higher the fixture, the greater the distribution and the more even the lighting. However, many municipalities restrict the permissible mounting height for luminaires not maintained by the municipality. Refer to the municipality ordinances for further details on permitted mounting heights.

LIGHTING LAYOUT AND DESIGN

Lighting design has been dramatically improved with the development of design software. Lighting designers can get an accurate representation of the lighting levels for a particular location based on luminaire placement, site design, and structure configuration. This section describes the process that designers use. It is based on the use of LitePro version 2.030. LitePro is a lighting design software package maintained by Columbia Lighting, Inc. For further details on the software, please refer to the developer's website.

Layout/Site Plan

Lighting a particular area requires a sense of the activities to be performed in an area and the dimensions of the area. For projects to be constructed, a building layout plan or site plan is required. The building layout plan is generally required for specifying interior lighting. The site plan is required for outdoor projects. Either of these documents will be developed for a project by an architect or engineer. These documents are usually essential as part of the lighting design phase.

LitePro is has the capability of importing Drawing eXchange Format (.DXF) files. The .DXF setup was developed by the Autodesk Corporation to facilitate drawing exchange between software applications. Layout and site plans generally contain a large amount of information that is not necessary for lighting design. Further, LitePro is more efficient importing drawings without extemporaneous details. As a result, it is suggested that the base layout or site plan drawing be modified by removing unnecessary information. These drawings will only be used for lighting layout so the original layout or site plans can remain.

Importing the drawings is accomplished by exporting the original drawing files to .DXF file type. When importing the drawings, LitePro needs the scale of the drawing. It is important that the correct scaling be used or calculations will be off by a factor of twelve.

LitePro also has the option of generating areas. This option is useful for simple layouts or sites or for projects where no drawings are available. The lighting designer simply specifies the dimensions of the area in the X and Y directions. For indoor areas, the ceiling height and slope must be specified. For outdoor areas, no ceiling height is required.

Luminaire Selection

Selection of the desired luminaire is essential for meeting lighting guidelines with the fewest fixtures possible and limiting light trespass while meeting local ordinances. LitePro is a product maintained by Columbia Lighting, Inc. and has a built in database of luminaires made by Columbia. The database is easily searchable depending upon several characteristics including indoor or outdoor use, light source type, fixture style, and quantity of lamps. Selection of the appropriate luminaire from the specified search criteria is done by trial and error or by the lighting designer's expertise.

In addition to the luminaires provided in the supplied data base, lighting designers can specify any type of luminaire by any manufacturer provided there is an .IES type file for the fixture. .IES files contain photometric (light output) data in a format standardized by the Illuminating Engineering Society. Without a lighting analysis application program such as LitePro, the files do not contain easily understandable information. To utilize an .IES file, the user simply imports the .IES file into LitePro and then provides a unique identifying number for the luminaire.

A common design activity for indoor or outdoor projects is to use a standard lighting fixture. For outdoor projects, a standard fixture with multiple luminaires mounted in different aiming rotations, tilts, and heights can be created in LitePro to simplify construction. This also ensures that the correct analysis is performed on the multiple luminaires fixture. A lighting designer must balance the costs of construction of additional fixtures with fewer luminaires as opposed to the energy costs of fewer fixtures with more luminaires. In general, a lighting designer should not specify further than two luminaires per fixture aimed in opposite directions, except for areas requiring higher levels of lighting than typical.

There may be a larger number of criteria for lighting a particular area or project. The obvious criteria are providing a minimum acceptable lighting level for a particular activity in a location. Some of the more subtle ones include locating lighting in the appropriate location. An example of this is using troffer style lights in a drop ceiling so that the troffers are placed in the actual drop ceiling grids when the lighting calculations are performed. Another is to place parking lot lights in locations where they will not impact traffic. These two criteria can be easily addressed in LitePro provided the layout/site plans have sufficient information to permit locating. Other criteria may be to provide a specific "feel" to lighting an area. An example may be a natural wood ceiling that uses up lights that provide a warming effect on the wood. This is more of a challenge to accomplish in LitePro and close coordination with the manufacturers producing up lights needs to be performed.

If a fixture is to be replicated at evenly spaced locations, such as a grid in a drop ceiling, multiples of the fixture may be placed on a specific area. The lighting designer simply specifies the number of grids in the X and Y directions and the spacing in each direction for the luminaire. Note that LitePro recognizes the multiple fixtures but treats them as a single fixture. As a result, all fixtures must be modified; a single fixture within the grid cannot be modified.

Light Level Calculation

Once a lighting designer has developed a preliminary lighting design, it is important to perform a lighting calculation to ensure the guidelines are met. LitePro needs to know where lighting calculations will be performed and a calculation grid must be developed for each area. The number of calculations to be performed is dependent on the grid spacing. Typically, outdoor projects would use a 10'X10' calculation grid spacing unless sufficient resolution is not achieved. The lighting designer must reduce the calculation grid spacing until the resolution is adequate. For indoor projects, it is generally 1'X1'. Again, if the resolution is not sufficient or the space is so large that calculations are taking an exceedingly long time, the typical calculation grid size can be adjusted.

The calculation grid elevation is placed as to where the lighting guidelines are specified. For example, for an indoor office, it is important that the desktop lighting level be appropriate to perform necessary activities. As a result, the calculation grid elevation would normally be 30 inches above the finished floor, the typical height of a desk. For outdoor projects, it is more important for pedestrians and drivers to see the surface or roadway. As a result, the elevation would be 0 inches above the finished grade.

In many projects, obstructions exist that prevent the penetration of lighting. For example, permanently installed cabinets in offices may block lighting. For outdoor projects, buildings or landscaping may block lighting. It is important to identify those obstructions in LitePro so calculations performed on the calculation grid are accurate.

Once a calculation grid is complete, an analysis is performed. This can be performed by each individual area, multiple areas, or for the entire project. The results of the analysis provide a minimum value, maximum value, and average value. Also, a watt per square foot number is provided. To ensure that the average and watts per square foot numbers are accurate, calculation points within obstructions that would prevent light penetration must be removed from the area.

As the designer adjusts lighting locations, the calculations are performed multiple times. Occasionally, it is necessary for the designer to obtain an image of the area lighting. This is accomplished in LitePro as a rendered image. The resolution of the rendered image is adjusted as necessary to show a higher quality image. High resolution renders may take an exceedingly long time and designers are encouraged to start with lower resolution attempts initially. Further, the rendered image may be exported from LitePro for presentation to owners or clients.

Once the lighting designer is satisfied with the final design, the designer will generally export the locations and descriptions of the fixtures so that they can be incorporated into the master layout/site plans. Incorporating into the layout/site plans ensures that the luminaires will be installed in the correct locations. This is accomplished in LitePro via an export to .DXF file format. The lighting designer needs to identify which items to export. Generally, it is important to export only the site plan, fixture location, and luminaire schedule. The site plan export is necessary to ensure that luminaires are referenced to the same objects in the master layout/site plans. For projects requiring municipality review of outdoor projects, it may be necessary to export the calculation grid with the lighting levels specified. This is accomplished in LitePro by toggling on the calculation grid for each area of interest. The lighting designer than copies the lighting information onto the site-plan from the .DXF file.

CONCLUSION

Artificial lighting has greatly expanded our capability and comfort. Artificial light promotes greater working efficiency by providing activity and task lighting that makes the activity easier to accomplish. Artificial light also makes items more appealing by providing a warming or calming effect. Artificial lighting can encourage traffic flow in a specific direction and can enhance safety. Artificial light can also create problems by over-lighting an area causing health issues.

Lighting design capability has been greatly enhanced by the introduction of lighting design software applications. Lighting applications, such as LitePro, are easy to use and effective at providing lighting designs that are efficient and meet design guidelines. Lighting guidelines are available from a variety of sources including the Illuminating Engineering Society. With good layout/site plans a lighting design can develop a well-lit project meeting all project criteria and guidelines while minimizing energy consumption.

References

1.   DiLaura, David, Houser, Kevin, Mistrick, Richard, Steffy, Gary, "*The Lighting Handbook, 10th Edition*," Illuminating Engineering Society, Copyright 2011.

Section II

Lighting Control
Considerations

Chapter 6

Review of Lighting Control Equipment and Applications

R.R. Verderber

Many types of lighting control equipment permit the automatic management of lighting systems. Lighting control equipment can be used to lower light levels, turn lights on and off on a schedule, and respond to the availability of natural light or the presence of occupants. Each of these operations minimizes the energy consumed while providing the proper illumination for use of the space.

The choice of a lighting control system depends on its particular application (retrofit, renovation, or new construction). This chapter will present a basis for selecting the control equipment and the control strategies (light reduction, scheduling, tuning, lumen depreciation, day lighting, load-shedding) most appropriate to an application, along with the important cost factors (initial, operating, installation, supply circuit layout, building design) that must be considered for each application.

LIGHT CONTROL EQUIPMENT

Table 6-1 lists various types of equipment that can be components of a lighting control system, with a description of the predominant characteristic of each type of equipment. Static equipment can alter light levels semipermanently. Dynamic equipment can alter light levels automatically over short intervals to correspond to the activities in a space. Different sets of components can be used to form various lighting control systems in order to accomplish different combinations of control strategies.

---

*This work was supported by the Assistant Secretary for Conservation and Renewable Energy, Office of Building Energy Research and Development, Buildings Equipment Division of the U.S. Department of Energy under Contract No. DEAC03-76SF00098.

Table 6-1. Lighting Control Equipment

System	Remarks
STATIC:	
Delamping	Method for reducing light level 50%.
Impedance Monitors	Method for reducing light level 30, 50%.
DYNAMIC:	
Light Controllers	
Switches/ Relays	Method for on-off switching of large banks of lamps.
Voltage/Phase Control	Method for controlling light level continuously 100 to 50%.
Solid-State Dimming Ballasts	Ballasts that operate fluorescent lamps efficiently and can dim them continuously (100 to 10%) with low voltage.
SENSORS:	
Clocks	System to regulate the illumination distribution as a function of time.
Personnel	Sensor that detects whether a space is occupied by sensing the motion of an occupant.
Photocell	Sensor that measures the illumination level of a designated area.
COMMUNICATION:	
Computer/Microprocessor	Method for automatically communicating instructions and/or input from sensors to commands to the light controllers.
Power-Line Carrier	Method for carrying information over existing power lines rather than dedicated hard-wired communication lines.

Static Controls. These controls can decrease light levels by a discrete amount, usually 30 or 50 percent. They provide a rapid, economical way to reduce energy consumption in an over-illuminated area. The method can be as simple as removing two lamps from a four-lamp fixture (delamping) or installing impedance-modifying lamps or devices. Some impedance-modifying devices must be placed in the fixture and hard-wired to the supply power, adding to installation costs.

Dynamic Controls. Automatic dynamic lighting control systems can consist of a combination of lighting devices such as controllers, which dim or switch lamps on and off; sensors, which measure light levels or sense the presence of occupants in a space; and communications, which process information from sensors and pass instruction to the light controllers.

LIGHT CONTROLLERS

Fluorescent lamps can be switched on and off via switches or relays. Because switches or relays must be hard-wired into power lines, they are most economical when they control one phase of a circuit or a large bank of lamps. Some devices can vary the amplitude of line voltage or duty cycle to standard core-coil ballasted fluorescent lamps. The light output of the lamps can then vary continuously from full to about 50 percent. Since they operate by conditioning the supply power, they are best suited to large banks of lamps. The systems are relatively economical to install, since they involve a central control. However, their flexibility is limited by the supply circuit layout.

Solid-state dimming ballasts are designed to continuously control the light output of fluorescent lamps down to 10 percent of full light output. The command to the ballast is via low-voltage signals (0 to 12 volts). Groups of lamps can connect on the same communication link with low-voltage wire. Thus, the flexibility of this method is not limited by the supply circuit layout. Each ballast has a manual adjustment so that maximum light output of each fixture can be achieved. This permits the use of all the lighting control strategies.

Some Control Devices

The sensors in a control system are used to obtain information about occupancy, time, or available daylight. Some sensors can relay commands directly to lighting controllers to operate lamps in a prescribed manner or they can pass information to a control processor. Personnel sensors determine the occupancy of a given space by detecting motion. They are most effective for spaces that have only one occupant, since the greater the number of occupants, the less chance that the space will be unoccupied.

Personnel sensors on the market are self-contained systems that include a sensor and a light controller. Because they must be hard-wired into the supply power, installation costs are involved. They are most effective when used by occupants who spend large portions of their time away from their stations or at stations that are intermittently used by several occupants. The controlled area must be occupied infrequently and unpredictably for personnel sensors to be most effective.

Photocells are used to measure the illumination levels in a space. If there is a change in the prescribed illumination level, i.e., because of lumen depreciation or day lighting, photocells signal the electric lights to maintain the prescribed level. The photocell outputs can be sent to a lighting controller that can alter the light levels or to a central processor. The photocell requires low-voltage wiring that creates some installation costs. Photocell lighting control systems that need to be hard-wired to the supply power incur additional installation costs.

Clocks provide instructions to a lighting system in real time. The equipment can be as simple as a spring-loaded switch that will turn the lights off at a prescribed time. It can also be incorporated in a control processor that will control the lighting system according to a prescribed daily routine for an entire year.

Communication

A single, central processor can store information and satisfy the communication needs for very large areas. The relatively high cost of a unit means it is generally not cost-effective for controlling a small amount of floor space. The computer/microprocessor is centrally located and involves relatively small installation costs.

A power-line carrier is a technically feasible device for transmitting input and output signals to and from a central processor. A carrier is particularly attractive in retrofit applications because it does not require distribution of signal wire. For other applications one must consider the relative costs and advantages of using a power-line carrier or distributing low-voltage signal wire.

LIGHTING CONTROL EQUIPMENT AND STRATEGIES

Table 6-2 lists the lighting control strategies that can make use of the various types of control equipment with notes on the strategies listed for some equipment. The number of strategies in which a control

Table 6-2. Lighting Control Equipment

| System | Application | | Strategy | | | | | |
| | | Light Reduction | Scheduling | | Tuning | Lumen Deprec. | Day-lighting | Load-Shedding |
			Predict.	Random				
STATIC								
Delamp	Retrofit	X			X			
Impedance Modifier	Retrofit	X			X			
DYNAMIC								
Light Controller Switch/Relay	Retrofit		X					X
	Renovation		X					X
	New Construction		X				X	
Voltage/Phase Control	Retrofit	X	X					
	New Construction		X			X	X	X
Solid-State Dim. Ballast	Renovation		X		X	X		X
	New Construction		X		X	X	X	X
Sensors Clocks	Retrofit		X					
	Ren./New Const.		X					
Personnel	Retrofit			X				
	Ren./New Const.			X				
Photocell	Retrofit					X	X	
	Ren./New Const.					X		
Communication Computer/Microprocessor	Retrofit		X			X		X
	Renovation		X			X		X
	New Construction		X				X	
Power-Line Carrier	Retrofit		X			X	X	X
	Ren./New Constr.		X		X			

is most effective depends on the application (Table 6-3). The following points out some of the features of the control equipment.

Static Controls. The static controls, delamping and impedance monitors, can be used to reduce the light levels throughout a space or in selected areas, i.e., by tuning. Since these strategies are most effective in spaces that are over-illuminated, they are best used in retrofit applications.

Dynamic Controls. The dynamic lighting controls include the use of various control components to create a system that executes the types of control strategies that are apropos depending on the application.

Light Controllers. In retrofit applications, relay-type controls are best limited to a single strategy, scheduling. A central processor with a clock can control an entire floor or building. If the power supply distribution can be altered, as in renovations, two control strategies can be accomplished. For appropriate buildings designed to exploit natural light, relay-type controls can employ three strategies (Table 6-2).

Voltage/Phase Control. These devices condition the input power to standard core-coil ballasts, which permits the dimming of fluorescent lamps over a continuous range. Because these devices dim lamps over a continuous range, these systems can accomplish two strategies in retrofit applications (see Table 6-2).

If photocells are used, three strategies can be accomplished, including lumen depreciation.

Table 6-3. Major Cost Factors for Lighting Control Applications

Application	Initial Cost	Operating Cost	Supply Installation Cost	Circuit Layout	Building Design
Retrofit	X	X	X	X	X
Renovation	X	X			X
New Construction	X	X			

In buildings designed to employ natural illumination, and with the proper wiring of the supply power, daylighting can be used.

Dimmable solid-state ballasts are best used in renovation and new construction because a ballast must be installed in each fixture.

In addition to the efficacious operation of the fluorescent lamp system, ballasts allow four major lighting control strategies. The advantages, as compared to the other controllers, are increased dimming range and the control of light levels via low-voltage signals. For buildings designed to exploit natural light, five strategies can be accomplished with solid-state dimming ballasts.

The strategies vary when using clocks, sensors and photocells. Clocks must indicate real time for employing the predictable scheduling strategy. Photocells are needed to measure ambient illumination levels for accomplishing lumen depreciation or the daylighting strategy. For lumen depreciation, only a few photocells are required. Considerably more photocells must be used for day lighting because the dynamic nature of natural illumination requires measuring illumination levels over smaller areas. Personnel sensors are most applicable to retrofit situations in spaces that are occupied intermittently during the day. They are the only devices that can automatically respond to unpredictable occupancy of a space.

Communications Controls. The communications central information processors and power-line carriers are auxiliary controls used in conjunction with lighting controllers and sensors.

They connect the instructions with the commands. The central processors are most effective with the scheduling, load shedding, and lumen depreciation strategies.

Transmitting information over power lines eliminates the need to string wire. In retrofit applications, rewiring could make the control technique too costly, so power-line communication is most attractive. In renovation and new construction, where rewiring will be done, power-line carrier methods are less attractive because of installation costs.

APPLICATIONS AND COST FACTORS

Decision-makers are faced with three types of applications choices (retrofit, renovation, or new construction) for a lighting control system. Table 6-3 shows the five major factors to be considered for each type of application.

A retrofit replaces or adds to an existing lighting system that is already operating adequately. The primary objective of a retrofit is to reduce operating costs. It is unlikely that a retrofit application would be economically sound if the supply circuit layout or the building design would have to be altered. The ideal retrofit requires no hard-wiring to the supply line. The above arguments are the reasons all five major cost factors must be considered for a retrofit.

When a building is renovated, the entire lighting system is replaced as well as all the supply circuit wiring. Thus, for renovation applications, installation costs and supply circuit layout are no longer major factors. That is, they would have been replaced at some cost in any case. The three major cost factors for a renovation are the initial and operating costs and the maintenance of the building design. This means the lighting control system can employ more control strategies since there are fewer application constraints.

New construction is an application where the lighting control system can influence the building design. For example, many new buildings are designed to optimize the use of natural illumination. The building structure, position, and fenestration system are designed to accommodate the lighting control technique.

Thus, the building design is not a major cost factor in new construction applications. This permits the use of day lighting. As shown in Table 6-2, a greater number of lighting control strategies are feasible in new construction.

APPLICATION AND EQUIPMENT

Table 6-4 lists the lighting control equipment that best suits each type of application. For each application, there are several options, which are optimum for different situations. The table lists control equipment in groups of priorities for each application. In addition the number of strategies that are optimum for each type of control are indicated in parenthesis. Priority I includes the equipment that meets all the conditions for the particular application. For example, delamping in a retrofit application has no initial cost, saves 50 percent in operating cost, has very small installation costs (removing the lamp and disconnecting the ballast from the supply), requires no change in supply circuit layout, and does not require a change in building design.

The group of retrofit equipment classified as Priority II involves significant installation costs and requires stringing wire throughout

Table 6-4. Equipment Selections for Various Applications

Priority	Retrofit		Renovation	New Construction
I	(1-2) Delamp	(2)	Switches/Relays	(2) Switches/Relays
	(1-2) Impedence-Modifier (lamps)	(3)	Voltage/Phase Control	(4) Voltage/Phase Control
	(1) Switches/Relays	(4)	Solid-State Dimming Ballasts	(5) Solid-State Dimming Ballasts
	(1-2) Voltage/Phase Control	(3)	Computer/Micro-processor	(4) Computer/Micro-processor
	(1) Computer/Micro-processor	(1)	Clocks	(1) Clocks
	(1) Power-Line Carrier	(1)	Photocells	(2) Photocells
	(1) Clocks			
II	(1-2) Impedance-Modifiers (hard-wired)	(5)	Power-Line Carrier	(5) Power-Line Carrier
	(1) Personnel Sensors	(1)	Personnel Sensors	(1) Personnel Sensors
	(1) Photocells			
	(3) Voltage/Phase Control			

( ) Numbers in parentheses indicate the number of lighting control strategies that can be implemented for an application.

the floor. These satisfy four of the five major cost factors.

The power-line carrier is in the Priority I group for a retrofit and Priority II in renovation and new construction applications. This is because laying control lines is not required when retrofitting. In renovation or new construction, the expense of laying the control lines is much less—that is, installation costs are not a major factor.

Personnel sensors are also rated Priority II for renovation and new construction. Although installation cost is not a major factor, the uncertainty of the activity in the space affects operating costs.

The difference in the number of strategies that can be accomplished in a renovated space and in new construction is shown in Tables 6-2 and 6-4. The additional strategy that can be employed in new construction is day lighting. This does not imply that day lighting is impossible in renovation applications. Daylighting depends on the building design, which is a major cost factor. However, some existing buildings are suitably designed to use daylighting effectively.

SUMMARY

A general approach for selecting the lighting control equipment that best suits an application is based on the interdependence of equipment, control strategies, major cost factors, and applications. The equipment listed for each group is not exclusive; there could be circumstances in which equipment or a strategy not listed would be suitable. More importantly, this chapter shows that a decision-maker has several options for lighting control systems for any application.

Chapter 7

Energy vs. Quality of Light:

How Lighting Controls Can Allow You to Optimize Energy Efficiency, Visual Comfort, and Space Performance

Eric Lind

OVERVIEW

Lighting accounts for more than 38% of electricity usage in commercial buildings,[1] resulting in more electricity being used for lighting than any other system.[2] Often, energy is wasted by lighting that is left on when a space is unoccupied, or when the area is over-lit by a combination of electric light and daylight. There are several strategies that can be used to reduce lighting electricity usage, saving a substantial amount of energy, while improving space performance and occupant comfort.

High-End Trim

High-end trim limits the maximum light output of fixtures. Because the human eye can hardly perceive the difference between 100% light level and 80% light level,[3] buildings can save energy without inhibiting occupant comfort or space performance. This results in potential energy savings of 20%,[4] while helping a building to qualify for LEED credits in the Energy and Atmosphere category.

Occupancy/Vacancy

Sensing occupancy sensors turn the lights on when a person enters a room and off when they exit. To increase energy savings, lights can be set to turn on to a lower light level, such as 75% as opposed to 100%, when a person enters a space. To save the most energy, occupancy sensors can be set to "vacancy" mode, which turns lights off when a space is vacated, but lights must be manually turned on upon entry. Because daylight is often available in rooms with windows or skylights, there is less of a need to turn lights on when a space becomes occupied. This strategy eliminates illuminating a space when the human eye can make its own determination of whether more light is needed. Occupancy sensors are ideal for stairwells (see Figure 7-1.) and restrooms where there are no windows, but vacancy sensors should be used in conference rooms, private offices, break rooms, and a wide variety of other spaces, resulting in energy savings of up to 15%.[5]

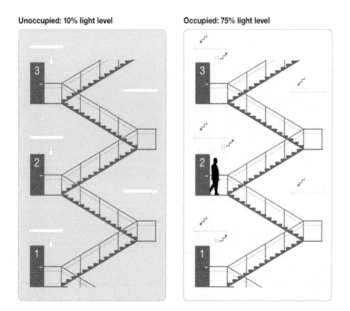

Figure 7-1. Stairwell Occupancy

Daylight Harvesting

If lights are left full-on during a bright day, a space can become over-illuminated, making it uncomfortable for occupants to perform day-to-day tasks. By installing daylight sensors to dim the fixtures closest to windows, you can take advantage of available daylight to save energy and enhance occupant comfort and productivity. Daylight harvesting contributes to LEED credits, complies with ASHRAE 90.1 2010 and IECC 2012 standards, and can save up to 15% of lighting electricity usage[6] in a space.

Personal Dimming Control

Research shows that people are more productive when working in their preferred light level. By providing the ability to adjust lights to perform the task at hand, you can increase productivity and decrease instances of absenteeism.[7] This strategy can save up to 10% of lighting electricity.[8]

Controllable Window Shading

By utilizing automated shading solutions, you can increase occupant productivity by decreasing glare and adjusting the amount of natural light entering a space. Shades can also block and reflect direct sunlight during warm days to reduce the demand on the building's air conditioning system, while providing an additional layer of insulation on cold nights (see Figure 7-2.). This yields up to a 10-30%[9] reduction in HVAC costs while meeting LEED guidelines for Energy and Atmosphere.

An often underappreciated aspect of this strategy is its value to ensuring savings from daylight harvesting. Without it, shades are often deployed and left lowered throughout the day. This neutralizes any savings available from reducing electric light due to available daylight.

Scheduling

Scheduling is an energy-saving strategy that dims or turns off lights based on the time of day or position of the sun. Lights can be

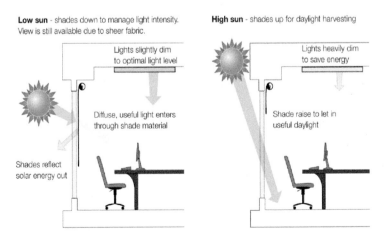

Figure 7-2. Effective Use of Automated Window Shading

programmed to turn off when buildings are often unoccupied, such as at night or during the weekend, which not only decreases electricity costs by up to 10%,[10] but it also reduces unnecessary light pollution.

Demand Response

During peak demand time, a utility company will contact a participating organization and request that they reduce their electricity consumption in order to decrease demand on the grid. Once an organization is contacted, they either manually or automatically reduce light levels, increase or decrease their temperature setting, adjust shades, and turn off non-crucial appliance loads. By participating, organizations can save money on energy during peak demand hours, while receiving incentives from utility companies for their efforts.

CONCLUSION

Incorporating light control strategies enables buildings to save energy and protect the environment by decreasing their carbon footprint, lowering greenhouse gas emissions, and reducing nighttime light pollution. These strategies also increase occupant comfort and productivity by supplying task-appropriate electric lighting, individual control, and glare reduction. All of this can be accomplished while saving money by increasing productivity and reducing costs.

References
1. Energy Information Administration, 2003 Commercial Buildings Energy Consumption Survey, released April 2009.
2. Energy Information Administration, 2003 Commercial Buildings Energy Consumption Survey, released September 2008.
3. Newsham, G.R.; Donnelly, C.; Mancini, S.; Marchand, R.G.; Lei, W. Charles, K.E.; Veitch, J.A. (August 13-18, 2008). The effect of ramps in temperature and electric light level on office occupants: a literature review and laboratory experiment. Institute for Research in Construction. Retrieved from http://www.nrc-cnrc.gc.ca/obj/irc/doc/pubs/nrcc41868/nrcc41868.pdf
4. California energy study. http://www.energy.ca.gov/efficiency/lighting/VOLUME01.PDF
5. IESNA 2000 Proceedings, Paper #43: An analysis of the energy and cost savings potential of occupancy sensors for commercial lighting systems. "Occupancy sensor savings range from 17% to 60% depending upon space type and time delay settings."
6. US Department of Energy. How to Select Lighting Controls for Offices and Public Buildings. Claim: 27% potential savings using daylight harvesting.

7. Determinants of Lighting Quality II by Newsham, G. and Veitch, J., 1996.
8. IESNA 2000 Proceedings, Paper #34: Occupant Use of Manual Lighting Controls in Private Offices. "Giving the occupant manual switching and dimming provided a total of 15% added savings above the 43% achieved by motion sensors."
9. Lutron commissioned simulation by Purdue University utilizing IES energy simulation software and real-world verification, November 2010.
10. When scheduling is used without occupancy sensing or vacancy sensing, 15% energy savings can be expected.

Section III

Reflectors

Chapter 8

Fluorescent Reflectors: The Main Considerations

Paul von Paumgartten

The benefits of replacing lamps in fluorescent light fixtures with reflector panels have been well documented. Articles have touched on everything from energy savings and reduced lamp and ballast maintenance expenses to lower air conditioning costs, extended lens life and reduced glare. These benefits have helped make reflector retrofitting the fastest growing segment of energy management.

At a glance, it appears simple enough. Investment in the retrofit and reap the return. But, in choosing to retrofit a building's fixtures, building owners and managers must closely examine a variety of considerations that will affect the budgetary savings, as well as the overall satisfaction of the building occupants. This chapter will spell out those technical factors that building managers and owners should evaluate when considering a fluorescent fixture retrofit plan.

Among the key considerations are:

- *Light Losses*. Replacing lamps with reflectors decreases light levels, but they may be maintained within current lighting standards and not adversely affect worker productivity.
- *Fixture Appearance*. The appearance of the fixture is critical, because people respond to lighting emotionally, as well as physically.
- *Designs*. The styles vary from full reflectors to partial, with varying effects on the lighting system.
- *Material*. Each has its own reflective qualities, affecting reflector output, appearance, and durability.
- *Costs*. The various reflective materials and designs present a cost-benefit dilemma.
- *Heating, Ventilating and Air Conditioning System Changes*. Fewer lights mean less heat, reduced cooling load and increased load.

LIGHT LOSSES

Logically, there is light lost when a lamp is removed and reflectors are installed. If two lamps from a two-by-four-foot, four-lamp troffer are replaced with optical reflectors, light losses may be 10 to 40 percent.

But, things are not always as they appear at first glance. Light losses vary from project to project depending on the fixture type and condition. Dirty lenses or old lamps cause fixtures to operate well below their designed efficiencies; they are dim to begin with. Buildings with poor lighting maintenance may be retrofit with reflectors with very little, if any, difference in light level. The illuminance that would be lost in an average reflector installation can be made up by cleaning the fixture and replacing the old lamps with new ones during the installation (see Figure 8-1). Installers can do this during a retrofit.

Fixture Efficiency

Fixture Condition	Efficiency
Seasoned and cleaned white troffer with new lamps and reflectors. (Retrofit fixture)	65-85%
Seasoned and cleaned white troffer with new lamps. (New fixture)	50-65%
Seasoned and uncleaned white troffer with seasoned lamps. (Old fixture)	35-45%

Figure 8-1. Reflector Charts

That's what makes the reflector concept work. Generally, a 10 to 25 percent light loss can be expected with a reflector retrofit.

In addition, the light loss may not be significant to the needs of the building occupants. Before the move to conserve energy, many building designers over specified for lighting. That may leave room to reduce lighting without cutting productivity.

As it is, new lighting standards suggest lower lighting levels are acceptable. Previously, most buildings were designed for light levels between 70 and 100 footcandles (FC). But, according to the Illuminating Engineering Society of North America lighting recommendations, offices need only 30 to 50 FC for Video Display Terminal usage; 50 to 70 FC for reading, writing, and typing; and 70 to 100 FC for accounting and drafting. In general lighting areas, offices need 10 to 20 FC for circulation areas, corridors and lobbies; 25 to 35 FC for conference rooms and non-task areas; and 30 to 40 FC for filing areas (see Figure 8-2). Task lighting can meet the needs of workers in many areas, while lower level ambient lighting is sufficient elsewhere.

Task Lighting Requirements	Footcandles
VDT Usage	30-50
Reading, writing, typing	50-70
Accounting, drafting	70-100
General Lighting Requirements	
Circulation areas, corridors, lobbies	10-20
Conference Rooms, non-task areas, work stations	25-35
Filing areas	30-40

Figure 8-2. Recommended Light Levels

Previously, building designers used 2.5 to 3 watts per square foot for lighting. Today, the standards have been reduced to 1.5 watts per square foot-without necessarily reducing occupant satisfaction or employee performance.

In addition to loss of light, reflectors can change the light distribution from a fixture. Reflectors direct light downward, creating more light directly under the fixture and less light between fixtures.

This light redirection is similar to that of the low-brightness louvers, which eliminate the high-angle light. This brightness control can reduce glare in offices where computer video display terminals are in use (see Figure 8-3). Spacing requirements may change because of less light between fixtures.

If the two-by-four-foot, four-lamp fixtures are providing more light than necessary, redesigning the fixture for reflectors could be readily acceptable to office workers. It is also less expensive than installing a new lighting fixture completely. But, several other key conditions remain in selecting the appropriate type of lighting designs using reflectors.

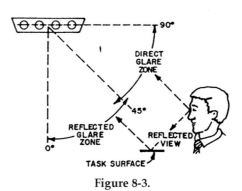

Figure 8-3.

FIXTURE APPEARANCE

Aesthetics, though difficult to quantify, are very important to a building environment. Many are obsessed with how much light is produced from a fluorescent reflector, but they may be over looking an important factor: the overall acceptance of a retrofit by the building occupants. Appearance can make a big difference in working conditions and worker satisfaction.

That is why delamping alone is not recommended for energy savings in an office setting. Delamping leaves a dark spot in the fixture where the lamps have been removed. This is aesthetically unacceptable to building occupants.

It has been shown that a great looking retrofit will more often prove to be successful than a bright one. Ambient lighting from overhead fixtures, task lighting for specific jobs or even accent lighting for visual affect must be done in an appropriate, tasteful manner.

DESIGNS

The reflector design can have an impact on aesthetics. As a minimum, a good looking retrofit fluorescent fixture should do one of two things: Create an image that the fixture contains a full complement of lamps or diffuse the light so no lamps can be seen in the fixture whatsoever. Different reflector shapes and materials will yield these results.

Some designs involve moving the lamp holders and repositioning the remaining lamp holders and repositioning the remaining lamps. Some virtually line the inside of the fixture with reflective materials. Others involve simply replacing existing lamps with reflectors. The reflectors may replace the in-board or the out-board lamps in a four-lamp fixture. Some have bends that help create the image of a lamp where one has been removed. And some are partial reflectors, which leave the ballast uncovered for easy access (see Figure 8-4).

MATERIALS

There are several different materials that can be used. Their reflective qualities differ, with silver film on aluminum rating the highest with a 90 to 95 percent reflexivity (see Figure 8-5). From a building occupant's

Standard troffer with 4 lamps.

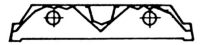

Complex bend reflector
with 2 lamps relocated.

Simple bend reflector with
2 lamps outboard.

Partial reflectors with 2
lamps outboard. Ballast
cover is not covered.

Complex bend reflector
with 2 lamps inboard.

Figure 8-4.

perspective, however, the difference between materials may be imperishable. Depending on the reflector design, the material used may have little effect on light output.

Figure 8-5. Material Reflectances

Material Types	Total Reflectivity
Silver film (on aluminum)	90-95%
Aluminum film (on aluminum)	80-85%
White paint (on aluminum)	75-90%
Anodized aluminum sheet	70-85%

Perhaps the best way to get a true impression of how a reflector design will look is to perform a mock-up. Have it installed in the ceiling to see the actual fixture where it will actually be used and measure the performance. Do not compare the measurements to the light levels that existed previously; instead, compare them to current lighting standards. The key question is: Will the reflector yield an acceptable light level?

Some types of reflectors will change in appearance over time. While aluminum reflectors have been proven durable in use for over 30 years, the thin anodized coating on aluminum reflectors—generally about two microns—may crack when the reflector is bent, causing possible oxidation of the aluminum.

While aluminum films have a long history, silver films have little track record. But research groups have run accelerated aging tests on silver films that show films with back reflector materials age faster, losing more than 20 percent of their reflectivity over 500 hours. Front reflectors maintained their reflectivity through the course of the study.

In time, temperature swings in the fixtures can also cause silver film to crack or buckle. Some film manufacturers, however, have added "ultra-violet inhibitors" to counter the effects of temperature swings.

Over time, a greater attention will have to be paid to lighting maintenance following a retrofit. Considerations should be made to group relamping on a regular, timely basis.

COSTS

While the effects of reflector designs and materials may have considerable impact on appearance, the retrofit's costs will have the greatest bottom-line impact. A high-cost retrofit could double the payback period.

The costs of the materials can vary greatly. Silver film reflectors are about $10 more expensive per fixture than aluminum. The average cost for a silver film reflector for a two-by-four-foot fluorescent fixture is $50 to $75 installed, compared to $40 to $65 for aluminum. Partial reflectors with aluminum film cost between $35 and $45 (see Figure 8-6).

Reflector Type	Typical Installed Cost	Payback*
Silver Film (Full)	$50-$75	2.3-3.4 yrs.
Anodized Aluminum (Full)	$40-$65	1.8-3.0 yrs.
Aluminum Film (Partial)	$35-$45	1.6-2.0 yrs.

*Based on 90 watts saved times 3,500 hours/per times seven cents/kWh or $22.05/year.

Figure 8-6. Cost Comparisons

Installation accounts for 25 to 40 percent of the cost. But lowest installation price may not be the best option, since proper installation is critical to the success of the retrofit. The installation contract should include new lamps and a thorough fixture cleaning to minimize the light loss in the retrofit.

The installation should also be done when it will least disrupt the building occupants. If performed after hours, the lamps can be removed and reflectors installed without the building occupants realizing there has been a change.

HVAC IMPLICATIONS

Improper installation or design can also mean additional hidden costs. Many light fixtures serve as return air ducts, an integral part of the heating, ventilating and air conditioning system. Full reflectors may cover the vents, impeding proper operation. Partial reflectors can render the necessary levels of illuminance without interfering with the air duct systems.

Also, building managers may not want to cover the ballast compartments. That increases the compartment temperature and shortens the life of the ballast. It also makes it more difficult to reach the ballast for maintenance procedures.

There are several considerations in HVAC systems that can drastically affect a building operation expenses as well as occupant comfort.

Reduced lighting causes a corresponding reduction in the cooling load for the air conditioning equipment, especially in the interior zones, where outdoor conditions have little influence. Demand for winter space heating may increase incrementally with reduced building lighting. This decreases the savings from the light reduction program by the amount of energy that must be added to offset the loss of heat. However, a building's heating system generally provides spare heat much more efficiently than the heat given off by excessive lighting.

For example, in a terminal reheat system, a change in lighting could require as much additional energy to reheat the duct air as is saved by reducing the lighting. The reheat requirement, however, can be minimized by raising the cool supply air temperature so comfort conditions in the room with the maximum cooling load are satisfied without reheating the air going to other rooms.

In the variable air volume system, a reduced cooling load would reduce the amount of cool air that is distributed through the building. This reduction may present an opportunity to replace the supply fan motors with smaller motors, saving additional energy. An HVAC expert is necessary to evaluate the retrofit savings potential.

A reflector retrofit also presents an opportunity to change the entire lighting system. Different lamps for better color rendition or greater efficiency could be installed. Parabolic louvres replacing the old lenses would reduce glare on office Video Display Terminals. And high-efficiency ballasts would afford even greater energy savings.

SUMMARY

With lights burning between 30 and 50 percent of a building's electrical load, many building owners and managers have chosen to install reflectors in their fluorescent fixtures as a quick-fix way to cut energy expenses. But the decision to install reflectors should involve more than simply changing a fixture. Aesthetics, designs, materials, costs and the retrofit implication to the HVAC system all need to be considered carefully. It also gives building management an opportunity to redesign the entire lighting system.

Careful selection of a reflector system and installer can help building management meet its desires for energy savings while increasing the satisfaction and productivity of building occupants.

Chapter 9

Specular Retrofit Reflectors for Fluorescent Troffers

J.L. Lindsey

INTRODUCTION

Specular optical reflectors for fluorescent fixtures can, according to some manufacturers' claims, permit the removal of 2 lamps from a 4-lamp troffer and actually result in an increase in lighting levels. Other manufacturers state that, in general, a reduction in maintained illuminance will occur but that the reduction is acceptable in many cases. These conflicting claims have generated substantial interest among consumers and members of the lighting community. Who is right and who is wrong? Many questions are asked, and the issue is clouded by lack of reliable information or accurate, detailed analyses.

This chapter will explore the performance characteristics and economics of several reflectors designed for use in prismatic lensed troffers, and their ongoing effects upon the performance of the lighting systems into which they are incorporated. Reflectors for troffers employing louvers, and for strip fixtures commonly used in industrial and warehousing applications, are not included in this analysis.

The photometric data used for the analysis were prepared by recognized independent photometric testing laboratories and provided by reflector manufacturers. Ballast factors are based on lamp manufacturers' data and accepted industry averages. Other light loss factors are obtained from the IES Lighting Handbook, 6th Edition, and from reliable industry sources.

Current typical costs for reflectors, lamps, luminaires, and labor were obtained from reflector and lighting distributors.

TYPES OF REFLECTORS

Reflectors are available in two basic types: semi-rigid reflectors which are secured in the fixture by mechanical means, and adhesive films which are applied directly to the interior surfaces of the fixture. Either silver or aluminum may be used as the reflecting media.

Silver reflectors are made by coating or impregnating a polyester film with elemental silver. The film may be bonded to an aluminum substrate to produce a semi-rigid silvered reflector, or coated with adhesive and applied directly to the fixture.

Semi-rigid aluminum reflectors are made from highly specular anodized aluminum sheet. Aluminized film reflectors are similar to silver films, but have lower reflectance.

Silver reflectors typically have reflectances in the 90% to 97% range, while aluminum reflectors range in reflectance from 70% to 90%. For comparison, white enamel paints used in most older fixtures and many current fixtures typically have reflectances of about 80% to 85%. Modem white powder paints generally exhibit reflectances of about 90%.

Reflector Design

Semi-rigid reflectors are bent to some specific configuration to direct light at desired angles, Figure 9-1.

The shape will be determined by the fixture construction, lamp placement, and desired lighting effect. Some reflectors are designed to maximize the amount of light which falls on the workplane by directing most of the light nearly straight downward. These designs may produce high lighting levels directly below the fixture, but areas in between fixtures may be somewhat darker as a result. Other designs direct some light outward at higher angles to provide better uniformity, and to lighten up wall surfaces. The quality of illumination produced by reflectors with wider beam spreads will normally be higher than more concentrating

Figure 9-1. Two of the many shapes used for reflectors.
Simple shape on left reflects light downward.
Complex shape on right creates multiple lamp
images and has a wider light distribution.

designs; however, the wider distribution units typically exhibit lower efficiencies. Dark walls and areas of low illuminance on the workplane may still be created if fixture spacing is at or near maximum for the un-reflectored fixture.

The design of the reflector is critical to performance. Analyses of photometric reports indicate that a properly configured reflector can be expected to improve fixture efficiency by 5% to 15% over a comparable two-lamp fixture without a reflector, or 17% to 27% when compared to a standard four-lamp fixture.

A poorly designed reflector can actually reduce efficiency to less than the efficiency of the fixture without a reflector.

Given comparable good design, a silver reflector can be expected to outperform an aluminum product by about 10%. If the silver reflector is poorly designed, a well-designed aluminum reflector will provide superior performance.

Film reflectors which are applied directly to the interior surfaces of the fixture are generally less efficient than semi-rigid reflectors since they conform to the fixture contours and cannot be formed to direct light in any specific manner.

WHAT HAPPENS
WHEN REFLECTORS ARE INSTALLED

The installation of reflectors in four-lamp troffers is generally accompanied by the removal of all four existing lamps, the installation of two new standard 40 watt lamps, and cleaning of the lens. Each of these actions has an effect on the performance of the fixture, and will occur independent of the installation of the reflector.

The Reflector

Reflectors increase the percentage of lamp lumens which reach the workplane. This is accomplished by redirecting light which is normally emitted at high angles, and the fact that reflectors generally reflect more light than a painted surface. The redirected light travels more nearly downward, so more light travels directly to the workplane, and less light strikes the walls. As previously stated, the redirection of light may result in darkened walls and areas of low illuminance between fixtures.

Effects of Lamp Removal

The lumen output of individual lamps and the percentage of lamp lumens exiting the fixture will both increase when two lamps are removed from a four-lamp fixture. These effects will occur due to a reduction in temperature within the fixtures, and the removal of mass which interferes with the exitance of light from the fixture. They occur simply due to lamp removal, and are not the result of a reflector installation.

Thermal effects result from the operation of fluorescent lamps at other than design temperatures. Standard four-foot lamps are designed to produce rated light output at bulbwall temperatures of about 100°F, which occurs in a still-air ambient temperature of 77°F for standard lamps and about 85°F for reduced wattage energy-saving lamps. Operation at higher temperatures will result in decreased light output, as shown in Figure 9-2a[1]. When bulbwall temperatures are reduced, light output also increases, as does wattage consumed by the lamps, as shown in Figure 9-2b[2].

Bulbwall temperatures in four-lamp enclosed troffers are difficult to predict, but research[3,4] indicates that temperatures of 120° F to 130°F can be expected. This results in a decrease in light output of 6.5% to 14%. The removal of two lamps can be expected to reduce bulbwall temperatures by 16°F to 19°F, and the loss of light due to temperature will drop to 0.5% to 2.5%, resulting in an increase in light output of 6% to 11.5%.

Replacement of
Energy-Saving Lamps with Standard Lamps

Energy-saving lamps have been installed in many four-lamp fixtures as an energy conservation measure. When standard lamps are used as replacements, the actual percentage of rated lumens which each lamp produces will increase since most ballasts are designed to operate standard lamps. Energy-saving lamps have slightly different electrical characteristics and, as a result, are not driven as efficiently by the ballast.

In general, standard lamps can be expected to produce about 94% rated light output, and energy-saving lamps about 87% rated light output when operated on commercially available ballasts. When energy-saving lamps are replaced by standard lamps, light output of individual lamps can be expected to increase by about 21% when replacing conventional phosphored lamps, and about 16% when replacing lite white lamps.

This means that two standard lamps will produce about 60% of the total lumens produced by four energy-saving lamps using conventional

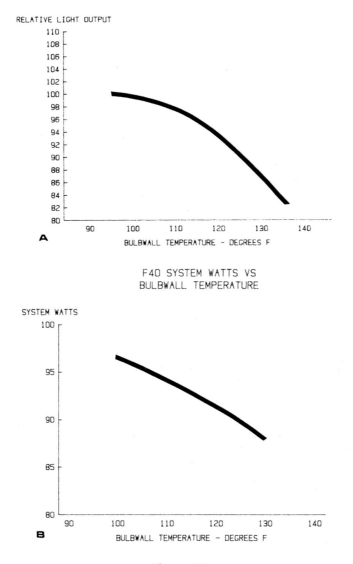

Figure 9-2
Thermal effects on light output and wattage consumed by standard F40 lamps.
Graph A shows light output as a function of bulbwall temperature. Graph B
shows watts consumed by 2 lamps and standard ballast with respect to temperature.

phosphors, and about 58% of the light obtained from four lite white lamps, simply because the standard lamps are driven more efficiently by the ballast. The percentages may be slightly lower under actual field conditions due to thermal factors.

Replacement of Old Lamps with New Lamps

As fluorescent lamps age they slowly decrease in light output. This is due to a gradual deterioration of the phosphors, and the deposition of evaporated cathode material on the bulb wall. At end of rated life, a standard phosphor lamp will produce about 76% of its initial lumens. Replacing an old lamp with a new lamp will recover this loss.

A typical lumen maintenance curve[5] for 40-watt rapid-start lamps is shown in Figure 9-3.

Effect of Cleaning the Fixture

The gradual accumulation of dirt on luminaire surfaces can have a profound effect on the performance of a lighting system. Lighting systems are seldom cleaned on a regular basis, yet routine maintenance can be one of the most cost effective ways of reducing lighting costs.

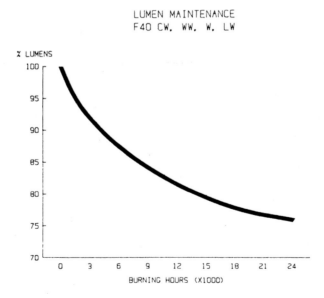

Figure 9-3. Lumen output of F40 lamps as a function of accumulated burning hours.

Reflectors are frequently installed in fixtures which have been in service for 5 to 10 years, yet have never been washed. Even in a clean office environment, the loss of light output due to dirt build-up in an unmaintained fixture can be as much as 35%, as shown in Figure 9-4.[6] Simply washing the fixture will recover this loss unless the fixture surface has deteriorate due to lack of cleaning.

When reflectors are installed, the lens is generally cleaned. The reflector is new and clean, and covers the dirty surfaces of the fixture. These actions recover the loss of light due to dirt accumulation on the fixture. The losses may also be recovered through fixture washing, either totally if the reflective properties of the fixture have not been impaired due to lack of maintenance, or at least partially if some deterioration has occurred. Permanent deterioration of luminaire surfaces is difficult to predict, and field evaluation or laboratory testing may be necessary to determine the extent of the problem. Reflectors, or even new fixtures, may be indicated if substantial deterioration has occurred.

Some reflector manufacturers claim that dirt does not adhere to reflector surfaces, and that reflectors are not subject to the same light loss factors as painted fixtures. Supporting data for this claim are not

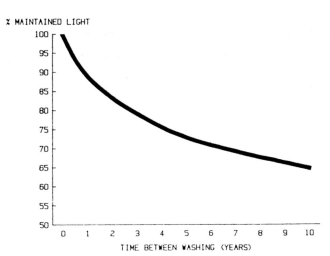

Figure 9-4. Expected loss of light due to dirt build-up for enclosed troffers in a clean office environment.

currently available. It appears reasonable, however, to assume that any reduction in light loss will be of small magnitude. An enclosed troffer has four surfaces which may gather dirt: the exterior surface of the lens; the interior surface of the lens; the lamp surface; and the interior surface of the fixture. The distribution of loss over these surfaces is not known; however, only one of the four surfaces is covered by a reflector. It can also be argued that dirt entering the fixture will settle on some surface; and if it does not settle on the reflector, it will probably settle on the lens.

Multiple Lamp Imaging

Reflectors can create multiple lamp images; a lamp may appear to exist where, in fact, it does not. Multiple imaging creates the visual appearance of having four or more lamps in a fixture which in reality has only two lamps. In some cases this feature may be the sole justification for installing reflectors instead of simply delamping. If aesthetics are a primary consideration, and the appearance of delamped fixture is not acceptable, reflectors may provide the solution.

Alternatives to Consider

When considering reflectors there are sufficient alternatives to fill a small book if each alternative is to be thoroughly analyzed. In addition to the options of various reflectors or simple delamping, there are high efficiency lamps and lamp/ballast combinations which might offer more attractive solutions. For example, if the performance of a delamped fixture without a reflector is marginally acceptable, but slightly more light is desired, high lumen output lamps might provide the answer at a considerably lower cost than reflectors.

The purpose of this chapter is to analyze reflectors and simple delamping; therefore, the other alternatives will not be discussed in detail. They may, however, provide a more viable means of reducing lighting cost, and should be considered.

There are seven basic options which should be evaluated when considering delamping in conjunction with the installation of reflectors in four-lamp troffers:

1.   Install silver semi-rigid reflectors.
2.   Install aluminum semi-rigid reflectors.
3.   Install silver reflective film.
4.   Install aluminum reflective film.

5.   Delamp without a reflector.
6.   Install new two-lamp fixtures.
7.   Do nothing, and continue with the existing system.

PERFORMANCE ANALYSIS OF ALTERNATIVES

The average illuminance produced by a lighting system may be predicted from:

$$FC = \frac{(\#\ lamps/luminaire)\ (Lumens/lamp)\ (CU)\ (LLF)}{Area\ per\ luminaire}$$

Where:
    FC   = footcandle
    CU   = coefficient of utilization
    LLF   = total light loss factor

If photometric data are available for the options which are under consideration, a site specific analysis may be prepared using this method.

The following general case analyses may be useful for preliminary evaluation of the alternatives, provided the parameters resemble conditions at the site. For purposes of analysis, the base assumptions are:

1.   The installation is in a 10-year-old office building which operates 12 hours per day, 5-1/2 days per week. The Room Cavity Ratio is 1.0, and the room surface reflectances are 80% ceiling, 50% wall, and 20% floor.

2.   The system was originally designed for 100 footcandles (maintained) using standard F40CW lamps with fixtures on 8' x 8' spacing for the first analysis. The second analysis is based on an alternate layout of fixtures on 6' x 8' spacing which produces about 125 footcandles maintained.

3.   Dirt condition is clean.

4.   Ballast factor is 0.94 for standard lamps and 0.87 for energy-saving lamps.

5.   Fixtures will not be washed and lamps will be replaced only upon burnout for the "Not Maintained" analysis.

6.   Fixtures will be washed every 12 months and group relamped every 3 years for the "Maintained" analysis.

Eight different scenarios, each with two different maintenance schedules, are analyzed for initial and annual end-of-year illuminance over a 10-year time span (Figures 9-5 and 9-6). Note that the performance of adhesive film reflectors is not included due to a lack of reliable photometric data. It is assumed that films will provide performance somewhere in between semi-rigid reflectors and no reflectors.

The column headed "No-Maint." assumes that luminaires are not washed and lamps are replaced as they burn out. The column headed "Maint." assumes that luminaires are washed on an annual basis and lamps are group replaced every 3 years.

Condition 1 represents the base case, an installation consisting of four-lamp fixtures with standard lamps.

Condition 2 uses the same fixtures but assumes that energy-saving lamps are used. This case is typical of many existing buildings where energy-saving lamps have been retrofitted as an energy conservation measure.

Condition 3 represents the most efficient silver reflector for which photometric data are available. A concentrating light distribution is used to direct light downward to maximize coefficients of utilization.

Condition 4 is the least efficient silver reflector found, and is used for comparison. Most silver reflectors will perform somewhere in between conditions 3 and 4.

Condition 5 is similar to condition 3 except an aluminum reflector is used. Reliable photometric data are not available for this configuration and a coefficient of utilization of 0.83 is assumed, based on a comparison of other photometric tests of aluminum and silver reflectors. While not precise, the estimated CU is believed to be sufficiently accurate for purposes of comparison.

Condition 6 represents the least efficient aluminum reflector for which photometric data are available. As with silver products, most aluminum reflectors can be expected to perform somewhere in between the extremes of conditions 5 and 6.

Condition 7 represents a system using typical new unmodified two-lamp fixtures. The performance characteristics of this system are similar to the aluminum reflector in condition 6, and the average illuminance produced will also be similar.

End of Year Footcandles
Typical 2' x 4' Troffer in Clean Office Environment
Luminaire Spacing: 8' x 8' Grid Pattern 64 sq. ft/fixt.
Ballast Factor: 0.94 STD Lamps, 0.87 Energy-Saving Lamps

	Condition 1		Condition 2		Condition 3		Condition 4	
	4 STD CW Lamps No Reflector CU = 0.69		4 E/S CW Lamps No Reflector CU = 0.73		2 STD White Lamps Silver Reflector #1 CU = 0.91		2 STD White Lamps Silver Reflector #2 CU = 0.76	
	Footcandles		Footcandles		Footcandles		Footcandles	
Year	No-Maint.	Maint.	No-Maint.	Maint.	No-Maint.	Maint.	No-Maint.	Maint.
0	128	128	111	111	84	84	70	70
1	103	103	90	90	68	68	57	57
2	92	98	80	85	61	64	51	54
3	86	94	75	82	57	62	47	52
4	81	103	70	90	53	68	44	57
5	77	98	67	85	51	64	43	54
6	75	94	66	82	50	62	42	52
7	73	103	64	90	48	68	40	57
8	71	98	62	85	47	64	39	54
9	69	94	60	82	46	62	38	52
10	68	103	59	90	45	68	37	57

	Condition 5		Condition 6		Condition 7		Condition 8	
	2 STD CW Lamps Alum. Reflector #1 CU = 0.83		2 E/S CW Lamps Alum. Reflector #2 CU = 0.73		2 STD WH Lamps New 2 Lamp Fixture CU = 0.73		2 STD CW Lamps Deteriorated Fixture CU = 0.60	
	Footcandles		Footcandles		Footcandles		Footcandles	
Year	No-Maint.	Maint.	No-Maint.	Maint.	No-Maint.	Maint.	No-Maint.	Maint.
0	77	77	68	68	68	68	64	64
1	62	62	55	55	55	55	52	52
2	55	59	49	52	49	52	46	49
3	52	57	45	50	45	50	43	47
4	48	62	43	55	43	55	40	52
5	47	59	41	52	41	52	39	49
6	45	57	40	50	40	50	38	47
7	44	62	39	55	39	55	37	52
8	43	59	38	52	38	52	36	49
9	42	57	37	50	37	50	35	47
10	41	62	36	55	36	55	34	52

Figure 9-5. Performance Analysis

End of Year Footcandles
Typical 2' x 4' Troffer in Clean Office Environment
Luminaire Spacing: 6' x 8' Grid Pattern 48 sq. ft/fixt.
Ballast Factor: 0.94 STD Lamps, 0.87 Energy-Saving Lamps

	Condition 1		Condition 2		Condition 3		Condition 4	
	4 STD CW Lamps No Reflector CU = 0.69		4 E/S CW Lamps No Reflector CU = 0.73		2 STD White Lamps Silver Reflector #1 CU = 0.91		2 STD White Lamps Silver Reflector #2 CU = 0.76	
	Footcandles		Footcandles		Footcandles		Footcandles	
Year	No-Maint.	Maint.	No-Maint.	Maint.	No-Maint.	Maint.	No-Maint.	Maint.
0	170	170	148	148	112	112	94	94
1	138	138	120	120	91	91	76	76
2	123	130	107	113	81	86	68	72
3	114	126	100	110	75	83	63	69
4	108	138	94	120	71	91	59	76
5	103	130	90	113	68	86	57	72
6	101	126	87	110	66	83	55	69
7	98	138	85	120	64	91	54	76
8	95	130	83	113	63	86	52	72
9	92	126	80	110	61	83	51	69
10	91	138	79	120	60	91	50	76

	Condition 5		Condition 6		Condition 7		Condition 8	
	2 STD CW Lamps Alum. Reflector #1 CU = 0.83		2 STD CW Lamps Alum. Reflector #2 CU = 0.73		2 STD CW Lamps New 2 Lamp Fixture CU = 0.73		2 STD CW Lamps Deteriorated Fixture CU = 0.60	
	Footcandles		Footcandles		Footcandles		Footcandles	
Year	No-Maint.	Maint.	No-Maint.	Maint.	No-Maint.	Maint.	No-Maint.	Maint.
0	102	102	90	90	90	90	85	85
1	83	83	73	73	73	73	52	79
2	74	78	65	69	65	69	46	65
3	69	76	61	67	61	67	43	63
4	65	83	57	73	57	73	40	69
5	62	78	55	69	55	69	39	65
6	60	76	53	67	53	67	38	63
7	59	83	52	73	52	73	37	69
8	57	78	50	69	50	69	36	65
9	55	76	49	67	49	67	35	63
10	55	83	48	73	48	73	45	69

Figure 9-6. Performance Analysis

Condition 8 represents a typical 10-year-old fixture which has been cleaned and delamped to two standard lamps. Failure to wash the fixture has resulted in permanent deterioration of reflective surfaces and a reduction in fixture efficiency. Note, however, that luminaire surface deterioration can vary widely between installations, depending upon the degree and severity of the contamination which caused the deterioration. This example assumes a reduction in efficiency of about 5%.

Uniformity of Illumination

The previous predictions of illuminance are average illuminance throughout the space. While they provide useful information, they tell nothing about the uniformity of illumination throughout the room, and undesirable areas of light and dark may exist. Uniformity is evaluated by measuring or calculating the illuminance at a series of points in the room. These points are usually on a grid pattern of sufficiently small scale to permit evaluation of changes in illuminance at points under and to the sides of luminaire locations, and at possible work station locations.

Figure 9-7 shows the results of these calculations for a series of points surrounding a single luminaire, as illustrated, in a room measuring 32' x 32', for the four luminaires used in the analysis of average illuminance. A luminaire spacing of 8' x 8' is used. Illuminance is calculated on a 1' x 1' grid pattern, and is approximately the same if transferred to other luminaires within the room. Note, however, that the illuminance in areas adjacent to walls will be slightly lower. These data are applicable only to the luminaires under study, and will vary if luminaires with different light distribution characteristics are used.

Figure 9-8 is similar to Figure 9-6 except that a luminaire spacing of 6' x 8' is assumed.

ECONOMIC ANALYSIS OF ALTERNATIVES

The following economic analyses compare silver semi-rigid reflectors, aluminum semi-rigid reflectors, silver adhesive film reflectors, new two-lamp fixtures, and cleaned existing fixtures which have been delamped, to an existing system consisting of unmaintained four-lamp fixtures.

87	89	91	94	94	94	92	90	89
88	90	93	95	96	96	94	91	90
91	93	96	99	100	99	97	94	93
95	96	100	103	104	103	101	98	96
96	98	102	105	105	105	102	99	98
94	96	100	103	104	103	101	97	96
91	92	96	98	99	99	96	94	93
87	89	92	94	95	94	92	90	89
85	87	90	92	93	92	90	88	87

A

54	55	59	63	64	63	60	56	55
54	56	60	64	66	65	61	57	55
56	58	63	69	71	69	64	59	57
57	60	66	73	76	73	67	61	59
58	61	67	74	78	75	68	62	59
57	60	66	73	75	73	66	61	58
55	58	63	68	70	68	63	58	56
54	55	60	64	65	64	60	56	55
53	54	58	62	63	62	58	55	54

B

C

49	51	55	58	60	59	55	51	50
50	52	56	60	61	60	56	52	51
52	54	58	63	65	63	59	54	53
54	56	61	66	68	66	61	56	54
54	56	62	67	70	68	62	57	55
53	55	61	66	68	66	61	56	54
51	53	58	63	65	63	58	54	52
49	51	55	59	61	59	56	52	50
48	50	54	58	59	58	54	51	49

D

48	48	49	50	50	50	50	49	49
48	49	50	51	51	51	50	50	49
50	51	52	53	53	53	52	51	51
52	53	54	55	55	55	54	53	53
52	53	55	56	56	56	55	54	54
52	52	54	55	55	55	54	53	53
50	50	51	52	53	53	52	51	51
48	48	49	50	50	50	50	49	49
47	47	48	49	49	49	49	48	48

Figure 9-7.

Illuminance at a series of points under and around a fixture for 8′ x 8′ luminaire spacing. The numbers in boxes represent the illuminance at points on the grid pattern within the room for several of the options. "A" is a typical unmodified four-lamp lensed troffer, "B" is a concentrating silver reflector, "C" represents a wider spread multiple imaging silver reflector, and "D" is an unmodified two-lamp fixture.

A

118	119	120	120	120	121	121
120	121	122	122	122	123	123
125	125	126	127	127	128	128
130	130	131	132	132	132	132
132	132	133	134	134	134	134
129	130	131	131	131	132	132
124	125	126	126	126	127	127
118	119	120	121	121	121	121
116	116	117	118	118	118	118

B

77	78	78	77	77	78	79
79	81	80	79	78	79	81
84	86	85	82	81	83	85
88	91	89	86	85	86	90
90	93	91	87	86	88	92
88	90	89	86	84	86	89
83	85	84	82	81	82	85
78	80	79	78	77	78	80
76	77	77	76	75	76	77

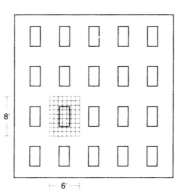

C

72	72	72	72	71	72	73
73	74	74	73	73	73	74
77	78	78	76	76	77	78
81	82	81	80	79	80	82
82	83	83	81	80	81	83
80	82	81	79	78	80	82
76	77	77	76	75	76	78
72	73	73	72	72	72	74
71	71	71	70	70	71	72

D

63	64	64	65	65	65	65
64	65	65	66	66	66	66
67	67	68	69	69	69	69
70	70	71	71	71	71	71
71	71	72	72	72	72	72
69	70	70	71	71	71	71
67	67	67	68	68	68	68
64	64	65	65	65	65	65
62	62	63	64	64	64	64

Figure 9-8.
Similar to Figure 9-6, except 6′ x 8′ luminaire spacing is used.

In order to evaluate the economics of maintaining approximately the same end-of-maintenance cycle illuminance for the alternatives, it is assumed that fixtures with silver semi-rigid reflectors will be washed and group relamped at 4-year intervals, fixtures with semi-rigid aluminum reflectors will be washed and group relamped at 3-year intervals, and fixtures with silver adhesive film reflectors will be washed every 18 months and relamped every 3 years.

The "new 2-lamp fixture" and "delamped existing fixture" options assume annual washing and 3-year lamp replacement.

Two different scenarios are presented; the first comparison (Figure 9-9) assumes that standard lamps are used in all systems. The second scenario (Figure 9-10) is based on energy-saving lamps in the existing system, and standard lamps in each of the alternatives.

The analyses assume a 20-year life for the systems, and cost of capital at 10%. These parameters will vary, but are assumed to be sufficiently accurate for a general case analysis.

Energy cost savings are addressed only from a lighting standpoint, and additional savings or costs attributable to air conditioning or heating are not included.

DISCUSSION

When considering any major lighting system modification, it must be remembered that the primary objective of the system is to provide illumination of adequate quantity and quality for the task being performed, at an acceptable (and frequently the lowest) cost. Aesthetics may also be an important factor, and must be evaluated with respect to the degree of aesthetics desired and the value they may have to the individual.

Before any conclusions can be reached about the suitability of reflectors for a given installation, the required quantity and quality of illuminance must be determined. Most buildings that are candidates for reflectors were built in an era when lighting systems were designed to produce 100 footcandles or more. In 1981, the Illuminating Engineering Society revised the recommended illuminance for most visual tasks, and established a method for selecting the quantity of light required.[7] Most office tasks are listed at 20 to 50 footcandles, with 30 footcandles applying to the typical task and worker. Some tasks. which are more visually demanding, are typically lighted to 75 to 150 footcandles; however, task-

BASIS OF ANALYSIS

This comparison evaluates 6 lighting systems on the basis of footcandles and cost. The lighting calculations are made in accordance with procedures established by the Illuminating Engineering Society. The cost evaluations are made using modern principles of life-cycle-cost benefit-analysis. Footcandles produced are the estimated minimum over a 10-year period, with the specified maintenance interval and assuming 8' x 8' luminaire spacing.

The luminaires evaluated in this study are:

Existing: Typical 4-lamp troffer   4   F40CW
No. 1 Existing fixture with silver reflector   2   F40CW
No. 2 Existing fixture with aluminum reflector   2   F40CW
No. 3 Existing fixture with silver adhesive film   2   F40CW
No. 4 Typical new 2-lamp troffer   2   F40CW
No. 5 Existing fixture, cleaned & 2 new lamps   2   F40CW

Bottom Lines	Existing Installation	Proposal No. 1	Proposal No. 2	Proposal No. 3	Proposal No. 4	Proposal No. 5
1. Footcandles produced	68	53	52	N/A	50	47
2. Investment needed	$ 0.00	48280.00	38280.00	31280.00	75280.00	5280.00
3. Capitalized cost	$ 0.00	46000.00	36000.00	29000.00	73000.00	3000.00
4. Total annual life cycle cost	$ 53631.21	32563.53	31678.27	31689.39	37690.95	29468.77
5. Annual savings	$ 0.00	21067.69	21952.94	21941.82	15940.26	24162.44
6. Payback period, years	0.00	2.12	1.66	1.39	3.85	0.23
7. Return on extra investment	0.00	43.64	57.35	70.15	21.17	457.62
Details						
8. No. of luminaires	1000.00	1000.00	1000.00	1000.00	1000.00	1000.00
9. Net cost per luminaire	$ 0.00	0.00	0.00	0.00	43.00	0.00
10. Reflector & labor cost/lum	$ 0.00	46.00	36.00	29.00	30.00	3.00
11. Net cost per lamp	$ 1.14	1.14	1.14	1.14	1.14	1.14
12. Lumens per lamp	3150.00	3150.00	3150.00	3150.00	3150.00	3150.00
13. Annual taxes & owning cost	$ 0.00	5403.15	4228.55	3406.33	8574.56	352.38

Figure 9-9. Economic Analysis (Cont'd)

	Existing Installation	Proposal No. 1	Proposal No. 2	Proposal No. 3	Proposal No. 4	Proposal No. 5
14. Annual power cost @ .08/kWh $	49695.36	25808.64	25808.64	25808.64	25808.64	25808.64
15. kW of load	181.00	94.00	94.00	94.00	94.00	94.00
16. Annual lamp & rep labor cost $	3935.85	811.18	807.75	807.75	807.75	807.75
17. Annual luminaire cleaning cost $	0.00	2500.00	2500.00	2500.00	2500.00	2500.00
18. Coefficient of utilization	0.69	0.91	0.83	N/A	0.73	0.69
19. Lamp lumen depreciation	0.82	0.82	0.84	N/A	0.84	0.84
20. Luminaire dirt depreciation	0.65	0.77	0.80	N/A	0.88	0.88
21. Ballast factor	0.94	0.94	0.94	N/A	0.94	0.94
22. Light loss factor — IES	0.50	0.59	0.63	N/A	0.69	0.69
23. Luminaire washing freq. mo.	100.00	48.00	36.00	18.00	12.00	12.00
24. Relamping interval, mo.	100.00	48.00	36.00	36.00	36.00	36.00

100 months means no luminaire washing or no group-relamping

Figure 9-10. Economic Analysis

BASIS OF ANALYSIS

This comparison evaluates 6 lighting systems on the basis of footcandles and cost. The lighting calculations are made in accordance with procedures established by the Illuminating Engineering Society. The cost evaluations are made using modern principles of life-cycle-cost benefit-analysis. Footcandles produced are the estimated minimum over a 10-year period, with the specified maintenance interval and assuming 8' x 8' luminaire spacing.

The luminaires evaluated in this study are:

Existing: Typical 4-lamp troffer	4	F40CW
No. 1 Existing fixture with silver reflector	2	F40CW
No. 2 Existing fixture with aluminum reflector	2	F40CW
No. 3 Existing fixture with silver adhesive film	2	F40CW
No. 4 Typical new 2-lamp troffer	2	F40CW
No. 5 Existing fixture, cleaned & 2 new lamps	2	F40CW

Bottom Lines	Existing Installation	Proposal No. 1	Proposal No. 2	Proposal No. 3	Proposal No. 4	Proposal No. 5
1. Footcandles produced	59	53	52	N/A	50	47
2. Investment needed $	0.00	48280.00	38280.00	31280.00	75280.00	5280.00
3. Capitalized cost $	0.00	46000.00	36000.00	29000.00	73000.00	3000.00

4. Total annual life cycle cost	$ 47109.52	32563.53	31678.27	31689.39	37690.95	29468.77
5. Annual savings	$ 0.00	14545.99	15431.24	15420.10	9418.57	17640.75
6. Payback period, years	0.00	2.91	2.28	1.91	5.69	0.32
7. Return on extra investment	0.00	30.13	40.31	49.30	12.51	334.10
Details						
8. No. of luminaires	1000.00	1000.00	1000.00	1000.00	1000.00	1000.00
9. Net cost per luminaire	$ 0.00	0.00	0.00	0.00	43.00	0.00
10. Reflector & labor cost/lum	$ 0.00	46.00	36.00	29.00	30.00	3.00
11. Net cost per lamp	$ 1.67	1.14	1.14	1.14	1.14	1.14
12. Lumens per lamp	2800.00	3150.00	3150.00	3150.00	3150.00	3150.00
13. Annual taxes & owning cost	$ 0.00	5403.15	4228.55	3406.33	8574.56	352.38
14. Annual power cost @ .08/kWh	$ 42831.36	25808.64	25808.64	25808.64	25808.64	25808.64
15. kW of load	156.00	94.00	94.00	94.00	94.00	94.00
16. Annual lamp & rep labor cost	$ 4278.16	811.18	807.75	807.75	807.75	807.75
17. Annual luminaire cleaning cost	$ 0.00	2500.00	2500.00	2500.00	2500.00	2500.00
18. Coefficient of utilization	0.73	0.91	0.83	N/A	0.73	0.69
19. Lamp lumen depreciation	0.82	0.82	0.84	N/A	0.84	0.84
20. Luminaire dirt depreciation	0.65	0.77	0.80	N/A	0.88	0.88
21. Ballast factor	0.87	0.94	0.94	N/A	0.94	0.94
22. Light loss factor – IES	0.46	0.59	0.63	N/A	0.69	0.69
23. Luminaire washing freq. mo.	100.00	48.00	36.00	18.00	12.00	12.00
24. Relamping interval, mo.	100.00	48.00	36.00	36.00	36.00	36.00

100 months means no luminaire washing or no group-relamping

oriented lighting is recommended for these higher levels. As a result of the revisions. an average design illuminance of about 50 footcandles is common. Option 5, simple delamping and maintaining an existing four-lamp system spaced on 8' centers presented in the study can be expected to produce over 50 footcandles for almost 2 years after washing and relamping, and drops to 47 footcandles at the end of a 3-year maintenance cycle. If the desired illuminance is 50 footcandles, the system would normally be considered acceptable. For more information on recommended illuminance and the illuminance selection procedure, the IES Lighting Handbook should be consulted.

Once the required illuminance is determined, a site-specific analysis should be prepared. The methodology used in this chapter may serve as a guideline.

For the general case analysis presented it appears that any of the alternatives presented could be acceptable for the typical office. provided that the existing luminaire has not deteriorated appreciably. This can be determined only by testing or a closely monitored field evaluation.

It is common practice to install one or several reflectors on a trial basis to permit field evaluations of performance in the client's facility. This may be a viable method of evaluating; however, the following procedure is recommended:

1.   Install new lamps and let them operate for 100 hours to burn in and stabilize.

2.   After the burn-in period, remove the lamps and thoroughly wash the fixture. lens, and lamps. Let the lens air dry. Do not dry it with a towel, since a static charge may build up, which will attract dirt.

3.   Re-install the lamps and let the system operate until the lamp bulb-wall temperature has stabilized. A period of three hours should be sufficient.

4.   Take illuminance measurements at a series of points under and around the fixture. The grid patterns shown in Figures 9-6 or 9-7 are recommended. Make sure that lamps have been removed or power turned off to all other fixtures in the area which might affect the test. Also make sure that no extraneous light from windows or other lights enters the area.

5.   Install the reflector and repeat steps 3 and 4. Mark the lamps before removal to assure that they are installed in the same sockets.

When evaluating the results of the test, remember that the measured illuminance is the initial illuminance produced by the system, and will deteriorate over time. If the system is washed annually and relamped at 10,000 burning-hour intervals (about 3 years in most large office buildings). the illuminance at the end of the third year should be about 75% of the initial illuminance.

Also be aware that there are differences between light meters. An inexpensive meter may have an accuracy of 10% or more, while precision photometers typically have 4% to 5% accuracy relating to the U.S. Standard, and 2% to 3% relative error. An inexpensive meter may read 45 or 55 footcandles, when in fact the illuminance is 50 footcandles. Be wary of placing absolute judgment on meter readings unless the characteristics of the meter are known.

Two areas of interest have been omitted from this discussion of reflectors: product durability and cost of washing. Aluminum has been used by the lighting industry for many years and has proven to be acceptable reflector material. Silver films are relatively new and their durability is somewhat unknown. There have been reported instances of film separation and flaking. and some films may have little abrasion resistance. When considering a silver reflector or film, check the manufacturer's track record. Obtain a sample of the material and test its abrasion resistance. Scratches in a specular surface will affect its reflectance characteristics, and result in a more diffused reflectance.

The cost used in the economic analysis for cleaning reflectors is assumed to be the same as an unreflectored fixture since cost data are not currently available. It appears reasonable to assume, however, that reflectors will be more difficult to clean than normal fixture surfaces, thus the cost will be higher.

CONCLUSIONS

From the data presented it appears that, in general, there are three instances where reflectors may be indicated in four-lamp lensed troffers: when the visual appearance of a delamped fixture is unacceptable; when the reflective surfaces of the existing fixture have deteriorated to an unacceptable level due to lack of maintenance; or when the original fixture has an unusually low efficiency, as is typically the case with narrow fixtures, which may be greatly improved by a redesigned reflector.

A comparison of silver and aluminum semi-rigid reflectors indicates a marginal improvement in performance with silver products. Given the higher cost of silver reflectors, it is doubtful that this improvement justifies the added cost. Performance characteristics of adhesive film reflectors are dependent on the fixture configuration; however, it appears that they will be most effective in shallow fixtures.

For the cases analyzed, reflectors may provide an expensive alternative to proper lighting system maintenance. In most cases washing intervals can be increased to three or four years with reflectors. instead of the annual washing which would typically be required if fixtures are simply delamped. The cost of fixture washing, however, is low in comparison to the cost of reflectors, and reflectors require a large initial investment. When the overall economics of reflectors is considered. it is difficult to justify their installation in most applications if simple delamping is aesthetically acceptable.

Most reflectors concentrate light under the fixture, and dark spots or areas of low illuminance may result. The uniformity of illumination should be evaluated to assure that adequate light is provided at workstations. and the uniformity ratios are not excessive.

An evaluation of the quality of illumination for tasks which are subject to veiling reflections, such as typical office work, should include an evaluation of Equivalent Sphere Illumination (ESI) for the various alternatives. The results of ESI calculations for three of the options analyzed in this chapter are shown in the following table:

| | Nadir | | Midpoint Between Fixt. | Across Dir. |
| | RAW FC | ESI FC* | RAW FC | ESI FC* |
Reflector				
Concentrating silver	78	25	58	57
Multi-imaging silver	70	21	54	52
2-lamp fixture no ref.	56	23	53	52

*North-viewing direction
From a qualitative standpoint, the three systems are approximately equal in terms of veiling reflections.

A decision to install reflectors, or any of the other alternatives to reflectors, should be made only after the evaluation of a site-specific analysis, which includes not only economic and quantitative factors, but also a consideration of the qualitative aspects of the illuminance produced by the lighting system. Due to the complexities of interactions between lighting system components, the analysis should be performed only by a qualified lighting professional.

References

1   Replotted from "Improved 35W Low Energy Lamp-Ballast System." Journal of the IES, April 1980. Illuminating Engineering Society, New York, NY
2   Ibid.
3   Ibid.
4   "Luminaire Retrofit Performance." Lighting Technologies, Inc., EPRI Project 2418-3, Electric Power Research Institute. Palo Alto, CA
5   Replotted from General Electric Company data on lamp lumen depreciation.
6   Replotted from IES Lighting Handbook, 1984 Reference Volume, Illuminating Engineering Society, New York, NY
7   IES Lighting Handbook, 1987 Applications Volume, Illuminating Engineering Society, New York. NY

Section IV

Ballast Selection

Chapter 10

Matching Fluorescent Lamps And Solid State Ballasts to Maximize Energy Savings

R.A. Tucker

ABSTRACT

Opportunities are available to maximize energy savings by using advanced energy efficient fluorescent lighting systems. This chapter discusses several systems on the market that benefit from recently introduced technologies.

Increased system efficiencies are possible by utilizing newly developed fluorescent lamps, electronic ballasts, and fixture types with improved optical characteristics. With proper combination of these advanced lighting technologies, an optimum design that provides significant reductions in both initial costs and system operating costs can be attained.

INTRODUCTION

During the past decade, energy consumed by lighting systems has been closely scrutinized for possible cost reduction consideration. For this reason, an assortment of lighting equipment and products were introduced by manufacturers to respond to the demand for increased efficiencies by users. The reason for this interest in lighting is simple. Lighting is important to the building owner because it can represent 50 to 60 percent of the commercial user's electric load. Therefore, from a business point of view, it is vital that we continue to look at lighting energy and. where possible. implement more effective and efficient advanced lighting technologies to cut operating expenses. The high cost of energy and the trend toward compulsory state regulations are focusing added attention on using lighting energy more efficiently.

California, Massachusetts and New York are examples of states with legislation limiting the watts allowed per square foot of space in a new commercial establishment. Legislation has been seen governing the power consumption limits on existing facilities and mandating a timetable to meet these lighting and HVAC limits.

In concert with these legislative actions. many utility companies across the United States offer rebates to commercial users to help fund the installation of energy-saving lamps. energy-saving ballasts, and energy-efficient lighting systems. They are spending large amounts of money to reduce demand and power consumption.

It is apparent that lighting energy management has become a priority. However, it still is an ignored opportunity by a majority of lighting users.

NEW LIGHTING SYSTEM DESIGN CONSIDERATIONS

Now that designers must operate in an area of heightened energy awareness and regulation, there is much interest in reduced diameter (T-8) fluorescent sources for both task lighting and general lighting applications. Matching these small diameter lamps with an electronic ballast results in higher lamp efficacies (lumens per watt) and increased system efficiencies. With the family of T-8 straight and U-shaped lamps increasing, additional design flexibility exists in terms of available wattages. The low operating current (265 $\underline{ma}$) for the T-8 family has permitted a lamp design compatible with a 60 Hz rapid start or electronic high frequency (25 kHz) instant start circuitry.

The highly efficient T-8 lamp-electric ballast combination has spurred the imaginations of many luminaire manufacturers as they develop an assortment of new fixtures. The better temperature performance of the T-8 design has allowed greater freedom in fixture design relative to optical and thermal considerations.

Consider the example of a 30-foot by 30-foot office space designed to be lighted by 2-foot by 4-foot parabolic louvered fixtures equipped with magnetic ballast and three 4-foot standard fluorescent lamps. The designer wants to maintain approximately 50 footcandles in this office with a ceiling height of 8.5 feet and the following surface reflectances: 80 percent ceiling, 50 percent walls, and 20 percent floor. Ten deep-cell parabolic 2-foot by 4-foot luminaires will be required representing 1.53

watts per square foot. The more aesthetic deep cell parabolic 2-foot by 2-foot fixture will require 15 units to do the same lighting job, representing 1.53 watts per square feet of lighting load (see Tables 10-1 and 10-2).

If we use an electronic ballast designed for the T-12 energy-saving 34-watt lamps in the 2-foot by 4-foot deep cell unit, then 10 fixtures will be needed to provide the 50 footcandles of illumination for a total of 1.0 watts per square foot. However, three reduced diameter T-8 U-lamps and the T-8 electronic ballast in a deep cell parabolic 2-foot by 2-foot system can outperform most other systems. In this example, the same space can be lighted to the desired illumination with 11 units, requiring only 0.94 watts per square foot. The only system that can outperform the T-8 U-shape lamp system is the T-8 straight lamp with the electronic ballast. It can light the space using 0.84 watts per square foot.

BALLASTS COMPATIBILITY

Like all fluorescent lamps. T-8 lamps must be operated on a ballast to limit the current and provide the required starting voltage. The ballast must be designed specifically for the lamp's electrical characteristics. the type of circuit on which it is operated. and the voltage and frequency of the power supply.

Table 10-1.

System Comparison for Various Lamp/Ballast Combinations

Fixture Type	Lamp Type	Ballast Type	Average Watts/Fixture	Average Watts/Sq. Ft.*
2x4 Parabolic 3-Lamp	T-12 Standard (40W)	Standard Magnetic	142	1.58
2x2 Parabolic 2-Lamp	T-12 U-Lamp (6-in. bend dia.)	Standard Magnetic	92	1.53
2x2 Parabolic 3-Lamp	T-8 U-Lamp	T-8 Magnetic	102	1.25
2x4 Parabolic 3-Lamp	T-12 Energy-Saving (34W)	Electronic	90	1.0
2x4 Parabolic 3-Lamp	T-8 Lamp (32W)	T-8 Magnetic	108	0.96
2x2 Parabolic 3-Lamp	T-8 U-Lamp	T-8 Electronic	77	0.94
2x4 Parabolic 3-Lamp	T-8 Lamp (32W)	T-8 Electronic	84	0.84

*To obtain 50 footcandles illumination in an office with 8.5-foot ceilings measuring 30-foot by 30-foot with surface reflectances of 80 percent ceiling, 50 percent walls and 20 percent floor.

Table 10-2.

System Comparison for Various U-Lamp/Ballast Combinations

Lamp Type	Ballast	Ballast[1] Factor	Watts[2]	Relative[3] Light Output (RLO)	RLO/W
Standard T-12 U-Lamp 2 Lamp 6-inch bend dia.	Standard Magnetic	.95	92	100	100
Standard T-12 U-Lamp 3 Lamp 3-inch bend dia.	Standard Magnetic	.95	145	128	81
Energy Saving T-12 U-Lamp 2 Lamp 6-inch bend dia.	Energy Saving Magnetic	.88	70	86	113
Energy Saving T-8 U-Lamp 3 Lamp	T-8 Magnetic, 3L	.95	102	138	125
Energy Saving[4] T-8 U-Lamp 3 Lamp	T-8 Electronic, 3L	.92	77	134	160

[1] Data in test normalized to ballast factors shown in this column for magnetic ballasts. Factors shown for electronic ballasts are measured values of sample.

[2] Data were obtained for a system using one-lamp and two-lamp ballasts. 3L designation refers to a three-lamp ballast system.

[3] Relative light output based on initial (100 hour) rated lamp lumen output.

[4] Life rated at 15,000 hours. All other systems shown are rated at 18,000 hours.

Because of its 265 ma. operating current, the T-8 lamp required the design of new rapid start ballasts. The magnetic rapid start ballast used in these lamps offers two primary advantages: (1) economical design and (2) long lamp life of 20,000 hours. Single and two-lamp magnetic ballasts are currently available for both 120 and 227 volt operation.

High frequency operation (20 kHz and up) improves fluorescent lamp efficiency, affording the opportunity to deliver the same light output for less power. Installations of high-frequency systems have been limited primarily by the cost and efficiency of the equipment required to convert power into higher frequencies.

Instant start circuits prove more economical at high frequency, because circuit design is simplified, with less complex ballasts and fixtures. The cathodes of T-8 lamps were designed with the knowledge that high-frequency operation would become more popular. Specific design considerations were incorporated into the lamps' cathodes to allow them to operate effectively either at high frequency or at 60Hz. Currently there are two-lamp, three-lamp and four-lamp high frequency instant start ballasts available. With these ballasts, lamps operate in parallel, so that if one lamp fails, the others continue to operate. All these high frequency electronic ballasts are also available in 120 and 277 volts.

While T-8 lamps operated on instant start have life ratings of 15,000 hours, they do provide the advantages mentioned above. The multiple lamp ballast designs simplify the equipment needed not only in four-lamp luminaires (by requiring just one ballast instead of two), but also in the increasingly popular three-lamp systems, which traditionally required both a single and a two-lamp ballast. Due to the increased efficiency of these systems, the shorter lamp life is more than offset by the reduction of power consumed to deliver the same light output. Lumen maintenance is improved to 91 percent at 40 percent of rated life when Octron lamps are operated on the high frequency instant start systems.

RETROFITTING THE EXISTING SYSTEM

An example of the potential for energy savings is the retrofitting of Tampa University using T-8 lamps and electronic ballasts. It began after extensive testing in a section of the university's library and will eventually extend to some 36 other buildings on campus situated on 69 acres along the Hillsborough River.

The Tampa Electric Company (TECO) initiated the energy reduction program through a monetary grant to the university for energy conservation and offered its expertise to the school. The library was targeted as a test site because its lighting is critical as well as relatively constant.

An area of the library known as the Florida Military Collection Room was selected for the test because the lighting circuits could be isolated for energy use measurement and documentation. Illumination levels recorded prior to the test were an average of 67 footcandles, including the contribution of daylight. The desired illuminance level for the room after relamping was set at a maximum of 100 footcandles. Relamping objectives were threefold: to meet the IES standard at the lowest cost, to provide a high color-rendering quality of light, and to reduce ultraviolet emissions that could harm the valuable paintings and artifacts in the room.

The room contained 42 three-lamp fluorescent fixtures and four four-lamp fixtures equipped with 35-watt cool-white fluorescent lamps.

An advantage for the T-8 system was its compatibility with existing bi-pin lamp holders in the 2-foot by 4-foot troffers. After evaluating various lamp/ballast systems (see Tables 11-3 and 11-4) and after meters were in place and a baseline established, 32-watt warm (3100K) T-8 lamps were installed. In addition, 46 electronic ballasts and new lens were installed.

Results immediately after the relamping and refurbishing showed a 38 percent reduction of 1.941 kilowatts, while illuminance levels had increased to 126 footcandles, more light than was needed in the room. Since the ballasts used could operate either three or two lamps, the number of lamps per fixture was reduced to two, for a resulting illuminance of 93.1 footcandles, not considering the contribution of daylight.

Based on the university's energy cost of 7 cents per kilowatt hour and the library's 4,365 hours of annual operation, the Collection Room and an additional 998 lamps in the library are projected to return all costs of the relamping, including new ballasts and maintenance, in less than 27 months.

Table 10-3.
Comparison of Four-Lamp Recessed Troffer Fluorescent Systems

Lamp Type	Ballast	Ballast[1] Factor	Watts	Relative[2] Light Output (RLO)	RLO/W
Standard T-12 , 40 watt	Standard Magnetic	.95	174	100	100
Energy Saving T-12, 34 watt	Energy Saving Magnetic	.88	139	91	114
Energy Saving T-12, 32 watt	Energy Saving Magnetic	.88	131	91	121
Energy Saving[3] T-12, 28 watt	Energy Saving Magnetic	.95	116	89	133
Energy Saving T-12, 34 watt	Electronic	.75	119	91	133
Energy Saving T-8, 32 watt	T-8 Magnetic	.95	032	101	133
Energy Saving[3] T-8, 32 watt	T-8 Electronic	.92	106	98	161

[1]Data in test normalized to ballast factors shown in this column for magnetic ballasts. Factors shown for electronic ballasts are measured.
[2]Relative light output based on initial (100 hour) rated lamp lumen output.
[3]Life rated at 15,000 hours. All other systems shown are rated at 20,000 hours.

Table 10-4.

Comparison of Three-Lamp Parabolic Louvered Fluorescent Systems

Lamp Type	Ballast	Ballast[1] Factor	Watts[2]	Relative[3] Light Output (RLO)	RLO/W[2]
Standard T-12	Standard Magnetic	.95	148 (139)	100	100 (107)
Energy Saving T-12, 34 watt	Energy Saving Magnetic	.88	115 (106)	90	116 (126)
Energy Saving T-12, 34 watt	Energy Saving Magnetic	.88	107 (99)	89	123 (133)
Energy Saving[4] T-12, 28 watt	Energy Saving	.95	102 (93)	86	125 (136)
Energy Saving T-12, 34 watt	Energy Saving Magnetic, 3L	.90	106	91	126
Energy Saving T-12, 32 watt	Energy Saving Magnetic, 3L	.90	95	87	134
Energy Saving T-12, 34 watt	Electronic, 3L	.88	90	90	148
Energy Saving T-8, 32 watt	T-8 Magnetic	.95	108 (104)	97	133 (138)
Energy Saving[4] T-8, 32 watt	T-8 Electronic, 3L	.93	84	96	170

[1] Data in test normalized to ballast factors shown in this column for magnetic ballasts. Factors shown for electronic ballasts are measured values of sample.

[2] Data were obtained for a system using one-lamp and two-lamp ballasts. Data in ( ) are for a tandem wired system using two-lamp ballasts. 3L designations refer to a three-lamp ballast system.

[3] Relative light output based on initial (100 hour) rated lamp lumen output.

[4] Life rated at 15,000 hours. All other systems shown are rated at 20,000 hours.

Section V

Natural Daylighting

Chapter 11

Natural Daylighting—
An Energy Analysis

RP. Jarrell

INTRODUCTION

As energy prices continue to rise, it is becoming imperative that our country's high rate of consumption be reduced. While new technologies are available, and are being developed to ease this consumption rate, significant reductions can be realized through the design and construction of energy efficient structures using existing technologies.

Building owners and developers must realize that they can no longer construct buildings with little or no concern for their energy consumption. If this practice continues, these owners and developers may find that their structures are too expensive to own and operate because energy prices have continued to rise.

Great reductions in a structure's energy consumption rate can be achieved with little change in its appearance or construction and, often, with little increase in the initial construction cost. Higher initial construction costs can be recovered in reduced operating costs.

With these factors in mind, an investigation into one aspect which affects the energy efficiency of a structure is presented. This study investigates the implications of using day lighting in a building and how it impacts the thermal load.

From this entire investigation, the author illustrates how architects have a great potential for providing energy efficient structures. Also highlighted are ways in which the energy performance of a structure can be greatly improved in the initial design phase. Finally, an analysis procedure is developed which will aid architects and others in the design of energy efficient structures.

PROCEDURE

For this investigation, an existing large-scale office building located in Atlanta, Georgia, was selected. The Georgia State Twin Office Towers project selected was designed in 1977 and was considered an energy efficient structure when completed.

The building's specific design features, construction methods and materials were determined. Daylighting availability levels were established. Other factors such as the building's mechanical and electrical equipment and energy management techniques were determined. Actual building usage data were developed through owner interviews and on-site investigations.

From these data, the energy consumption of the project was estimated. This was accomplished by performing a LOADCAL computer analysis on a representative portion of the building. A typical open plan office level was selected for analysis. The analysis accounted for specific building conditions, existing within the facility, as well as actual environmental factors.

Upon completing this analysis, the availability and usage of daylighting in the existing building was determined. This procedure involved estimating daylighting levels at multiple locations within the typical office level.

Then, alternative building schemes were analyzed to determine if a more efficient design could be provided. This revised scheme did not, however, significantly alter the present architectural design parameters of the facility.

With the objectives and methodology of this report outlined, a detailed description of the Twin Towers will be presented. The discussion will highlight the building's major construction features and materials. Electrical and mechanical systems are described along with energy saving and management techniques utilized in the design, construction and operation of the facility.

PROJECT DESCRIPTION

The Twin Towers office complex consists of four distinct interconnected facilities: the Georgia State MARTA station, the entry plaza, the office towers and the central energy plant. The project, located across

Martin Luther King Drive from the Capitol, is also bounded by Piedmont Avenue, Butler Street and by the Georgia Railroad trackage.

The MARTA Station is composed of a lower concourse, allowing entry from Piedmont and Butler, an elevated platform where the trains are boarded and an upper concourse which provides entry to the 'plaza' of the Twin Towers. The station is a totally ·independent facility in terms of operation, maintenance and security. The only link to the Twin Towers is to provide access to the rapid-rail system.

The Plaza is the structure which serves as the base beneath the office towers. (Within its four levels are maintenance shops, storage and mechanical space on the lowest level; a cafeteria and snack bar, post office and mechanical space on the second level.) The third level is the main public access level and contains office space for state agencies which require maximum public contact. The upper plaza level coincides with the upper concourse of the MARTA station and serves as the access point between the building and the MARTA system.

The two towers rest on the plaza base and each tower contains 16 typical office floors. Each floor is designed to accommodate both open and/or traditional office plan layout, but open office layouts are predominant. (For this study, a typical open plan office floor was selected.) A mechanical penthouse caps each tower.

The fourth facility is the Central Energy Plant. It is partially underground and located adjacent to the south of the towers structure. Provisions for landscaping and access walkways have been made on its roof and thus it sits under the 'front-yard' of the Twin Towers. The central energy plant provides chilled water and electrical service to the Twin Towers.

Brief History

The impetus for the construction of these facilities was the CAPITOL HILL 2000 plan, a long-term master plan for development of facilities in the vicinity of the state capitol. A key recommendation of the master plan was the construction of two symmetrical state office buildings to house many of the state agencies. The buildings were to be located in air-rights over the proposed rapid-transit station. This would afford both state employees and the general public easy access to these agencies and to Capitol Hill in general. At the same time, the office space provided would enable the state government to increase its efficiency by consolidating, into one location, many of the personnel who were located through the downtown area in leased office space.

In early 1975, the architect, Aeck Associates, Inc., was selected to prepare preliminary studies for the towers and its plaza. This task lasted into early 1976. Simultaneously, MARTA with its own consultants commenced designing the station which would interface with the office tower project. The preliminary drawings were completed for the Twin Towers in February 1976 and the final design phase began. This phase lasted through late 1977. Construction began soon thereafter and continued through November 1981. The building was occupied in phases and became fully occupied in June 1983.

The Building Concept

As stated, the architect was charged with designing a facility with two identical towers. In order to improve the constructability of each tower, the number of floors was reduced from the concept outlined in the CAPITOL HILL 2000 concept. Furthermore, the area of each floor was increased to improve the space utilization within each tower. This measure eliminated the need for two tiers of elevators, which afforded both economic and energy savings. Because of the basic premise of the planning concept, the two-tower scheme, no consideration was given to combining the towers into a single larger-scaled tower.

Architectural Considerations

The building skin consists of masonry and glass. Prefabricated, medium-color utility brick panels were selected as the predominant exterior cladding material. This selection was based upon a study of aesthetics, construction costs, life-cycle costs and thermal mass.

During the design, studies were made analyzing glazing types and amounts of glass areas. These studies were based on aesthetics, natural versus artificial light, HVAC loads and views to the outside. The amount of glass utilized in the final design, as a percentage of the building enclosure within a typical 25-foot-wide office bay, is approximately 39 percent. The glass on the north and south elevations is shaded by a brick spandrel which projects 18 inches. East-west facing glass is shaded by a IS-inch brick spandrel. Brick column surrounds, which serve as vertical fins, help to shade the glass with north-south projections of 20 inches and east-west projections of 34 inches.

The glazing selected is Solarban 550-60. It is a one-inch insulated, bronze-tinted glass manufactured by Pittsburgh Plate Glass. The following table describes the design data on the glass used in the Twin Towers.

Table 11-1. Glass Design Data

U-value (summer)	= .59
U-value (winter)	= .50
Shading Coefficient	= .46
Visible Light Transmission	= 32%

Various window framing solutions were analyzed including operable windows, provided with and without integral blinds. Operable windows, which appeared a common sense solution, were shown not only to be cost prohibitive (approximately $500,000 in additional costs) but were counter-productive from an energy use standpoint.

By having a fixed window frame, the possible introduction of latent heat from humid summer air is reduced. In winter, the possible loss of heat through exfiltration is reduced by eliminating the large amount of window perimeter 'crack' found with operable windows. Exterior airborne dust intake, which has adverse effects upon HVAC equipment, is also reduced.

Several types of insulation are used in the project. All construction U-values met those recommended by the codes in force at the time of the design and construction. The following table gives data on the major types found in the building.

Table 11-2. Insulation Data

Polyurethane and Perlite	
Roof Insulation	U-value = .06
Polystyrene Deck	
Insulation	U-value = .06
Rigid Fiberglass Masonry	
Panel Insulation	U-value = .11

Medium color, horizontal blinds have been provided throughout the entire project. The one-inch blinds have a positive stop at 70 percent of the fully closed position. The blinds reduce the amount of direct solar radiation gain within the spaces. It was pointed out by the mechanical engineer that the blinds reduce the window shading coefficient by, a maximum of, 28 percent to .33.

$$Shading\ Coefficient = .46.$$
$$46 \times (.28) = .1288$$
$$.46 - .1288 = .33$$

Mechanical equipment penthouses are located on the rooftop of each tower. The mechanical engineer noted that with a lapse rate of 3Y2 degrees (F) per 1000 feet, the location of the mechanical equipment on the roof affords the dominant cooling cycle a slight advantage over systems with a lower-level intake.

Mechanical Considerations

The engineer, Nottingham, Brook and Pennington, P.e., commenced by analyzing alternative fuel sources for the building. Coal, solid-waste, oil and natural gas were considered as was off-peak generation. Using the Trane Corporation's TRACE mechanical analysis program, energy analyses were performed evaluating eight options of HVAC equipment and energy sources. These analyses included such items as chiller and condenser equipment and fan equipment on variable volume systems. The building envelope was also analyzed using a computer program based on ASHRAE's Standard 90-75.

HVAC Features Incorporated: A variable-air-volume ventilation system is employed throughout the building. No 'reheat' is allowed within the system. According to the TRACE program, vane-axial supply and return air fans were selected which give the maximum fan power savings and the maximum turn-down ratio available.

Rotary total heat exchangers are used to capture approximately 72 percent of the loss/gain of the central toilet exhaust system. Because of the extended operation time of the building, this affords considerable savings over the course of a year. An air-side economizer cycle, with enthalpy control, reduces mechanical cooling requirements below 78 degrees (F) outdoor temperature. It also provides free cooling below 55 degrees (F) outdoor temperature.

Water wash electronic air filters are employed to provide a lower pressure-drop, thus reducing fan horsepower requirements. The filters selected have an efficiency rating of 95 percent.

Heating is provided by low-temperature, hot-water, finned piped convectors located under the perimeter windows of the building. This is the most efficient location and method for introducing heat into exterior spaces. The system is zoned by exposure to avoid overheating when solar radiation is available.

Space temperature and ventilation control is provided by induction boxes. These boxes, formally called 'heat-of-light' boxes, reclaim up to 50 percent of the heat from the lights by inducing warm air from the ceiling plenum to the exterior areas where heat is needed in winter. In the summer, this heat is returned to the mechanical penthouse where it is removed by the refrigeration system.

A two-way valve control system is employed on all chilled and heating water coils which provides reduced pump horsepower during lighter loads.

Boiler Plant: Two hot-water generators are incorporated into this facility. They are dual-fuel, '0' water-tube type with high turn-down ratios (10: I). The water tube boilers in this size (30 million Btu / hr were the most effective ones on the market at the time of construction. Because the boilers are set up on a dual-fuel capability, the owners have the opportunity to take advantage of synthetic fuels as they become economically available. They are presently set up to use natural gas and number two fuel-oil.

The hot water pumping system selected has variable speed capability to reduce pump power consumption during light loads. Heat reclaimed from the hot water pumps' mechanical seal cooling system is used to heat the boiler room and to preheat combustion air.

Control Systems and Automation: A central control room has been incorporated into the Twin Towers and from this point all systems within the towers are monitored. Critical functions are controlled by building engineers from this room.

The owners purchased a Johnson-Controls JC-80 building automation system in an effort to reduce energy consumption and to improve the control of the indoor environment. The towers were designed to incorporate the required conduit, to each floor, to interface mechanical control panels with the JC-80 system.

A number of energy conserving control measures are employed in the operation of the facility. Space temperature setbacks are employed during non-occupancy hours. Both the ventilation and the exhaust systems are automatically shut off during the pre-occupancy warm-up period. Automatic air-temperature and hot-water resets are employed based upon outside air temperature.

Electrical Considerations: Because occupancy requirements were not known prior to construction, considerable flexibility had to be designed into the lighting system. Both open and traditional office layouts were anticipated in the tenant occupancy. Therefore, a variety of conditions had to be addressed in the design of the lighting system.

Given the constraints cited above, four sources of illumination were analyzed using a life-cycle cost comparison. They are mercury, fluorescent, metal-halide and high-pressure sodium. Fluorescent lighting was selected based upon this analysis.

While comparisons of color, appearance and rendering properties were made, cost was the final determinant in selecting a four-tube fluorescent troffer system for general office illumination. The Fluorescent system provides the state's requirement of 70 footcandles, for standard office occupancy, with approximately 2.4 watts per square foot energy usage. This system provided a first-cost saving of about $407,000 over runner-up high-pressure sodium fixtures. The high pressure sodium system would only reduce the annual utility costs by approximately $27,500. The troffers installed are slotted to vent heat directly into the ceiling plenum.

High Intensity Discharge source fixtures have been provided in the lower levels of the plaza where greater ceiling heights exist. They are also used for general outdoor lighting. A minimum of incandescent lighting is used in seldom occupied spaces such as electrical and janitorial closets.

The owner undertook an energy conservation program to reduce the light-wattage, per square foot, in the system provided. In the open plan portions of each tower, two of the four lamps were disconnected in many in reception, waiting and circulation areas. The ballasts serving these delamped fixtures, however, were not disconnected. While the delamping of fixtures reduced the power consumption, greater reductions could be realized by disconnecting the unused ballast in each fixture. A final measure included individual switching of lights in individual offices. These steps were implemented under the direction of the project's design team.

Peak Shaving: Emergency generators for elevators and other electrical equipment are arranged so that they can be brought on-line as needed for electrical demand control. The central control system automatically shuts down selected mechanical and electrical equipment as a power conserving measure and operates the emergency generators to reduce peak electrical demands.

LOADCAL ANALYSIS

In the preceding section, items such as the building's major construction and materials, electrical and mechanical systems and the energy management systems were highlighted. These items were presented and discussed to highlight the systems and components used within the project. With this description provided we can now focus upon the analyses to be performed on the project. These analyses will be the basis for recommending any modification to the structure.

An energy analysis was performed on a representative portion of the building: a typical open plan office level. LOADCAL, the analysis program used, is based on the ASHRAE GRP-158 analysis procedure. It calculates the cooling and heating load of a space (or building). The program takes into account environmental and interior factors which affect the mechanical requirements of the space or building. To ensure that the results of this analysis were as representative as possible, an intermediate office floor was selected. This eliminated any effects caused by having a roof transfer load.

The typical open office floor was divided into four perimeter zones and one interior zone. Each perimeter zone has one exterior exposure. The orientation of the northern zone (zone four) is 40.82 degrees east of true north.

Floor, wall and glazing areas were taken directly from the architectural plans. Wall configurations, overhang and vertical fin dimensions were also derived from the plans. Data concerning U-values were established based on the information contained in the drawings and specifications. They were verified with the project's mechanical engineer.

Typical Atlanta weather data were used in this analysis. The data were taken from meteorological tables, with 1984 being the base year. The solar radiation intensity level, however, was set to equal 1.0 for all evaluations. This value represents a clear sky condition. Actual levels vary (of course) with the degree of cloud cover, level of smog and/or other atmospheric conditions which will block available solar radiation and, thus, reduce the intensity. The effect of using the 1.0 value is a somewhat over-estimated solar radiation load than found under normal sky conditions. The increased loading condition used for this analysis will be used in the revised building analysis, thus, allowing for comparable results. Actual intensity data for Atlanta is being developed by others, at the time of this writing.

Analysis data for interior areas represent actual building use conditions. This information was obtained through on-site inspections and from verification with the owners of the project, the Georgia Building Authority. Thermostats maintain a relatively constant year-round temperature of 74 degrees (F).

Humidity ratios used in this analysis are 0.0060 pound/pound for the winter months and 0.0 I 03 pound/pound in the summer months which represents typical Atlanta design data. It is the ratio of the mass of water vapor to the mass of dry air.

Light wattage data were calculated from the architectural and electrical drawings. They were verified through on-site visits and in discussions with the owner. Lighting levels, in footcandles, were checked using a hand-held General Electric illumination meter. The readings indicated that a minimum level of 70 footcandles existed throughout the office areas. Some areas had higher illumination levels.

The light-wattage value established for each zone represents lighting and equipment loads for that zone. Equipment loads were factored by the ballast factor, of the lighting equipment, to eliminate the possibility of over-estimating the total zone wattage value. Light fixture and lamp data were taken directly from the architectural drawings and specifications.

The ballast factor was obtained directly from General Electric, the supplier of the fixtures and the lamps. The delamping plan, as implemented by the owner, was accounted for in the determination of the lighting wattage data. The lighting usage data represent actual conditions, as the lights are left on continually from morning occupancy until the cleaning crews have finished late at night.

The building is typically occupied Monday through Friday, from 8:00 a.m. until approximately 5:00 p.m. The occupant loading is based upon the owner's design criteria of one person per 150 square feet of floor area. This closely represents actual conditions as state government buildings are occupied less densely than typical speculative office buildings. Zones one through four reflect this occupant load condition while zone five is a service zone and the occupant loading here, which is much less, was estimated.

The ventilation rate is ten (10) cubic feet per minute (per person) within the building. The mechanical system employs a rotary total heat exchanger, which is 72 percent efficient, thus the impact of the ventilation rate on the mechanical system is reduced as shown in the following equation.

Ventilation Rate = 10 cfm / person

Heat Exchanger = 72 percent efficient
    (1.0 - .72 = .28)

Energy Consumption from Fresh Air Induction =
    10 * .28 = .28 cfm / person

The system operates only while the building is occupied and therefore a "non-continuous" factor was entered into the LOADCAL program.

LOADCAL estimates the annual cooling and heating load to be approximately 1,307,394,251 Btu's, per year, for the existing office floor analyzed. This translates into approximately 64,088 Btu/sf/year for the typical, 20,400-square-foot, office level.

An appendix which contains the actual LOADCAL computer analysis of the representative portion of the facility will be available for review at the seminar. The input data, found on the initial pages, contain general project data, outside weather data and inside target data. Specific zone-by-zone data for all bays, derived to analyze the building, are also included.

Results from the LOADCAL analysis follow the input data and have been provided in a zone-by-zone format. A floor summary, which sums the results of the zones one through five, is also provided.

This concludes the LOADCAL energy analysis of the existing building. The following section estimates the available daylighting levels. This was accomplished using the methodology developed by Libbey-Owens-Ford.

DAYLIGHTING ANALYSIS

Analyzing the existing daylighting levels, within the typical office level, involved defining such factors as the typical level's geographical orientations and specific zone configurations. In the LOADCAL analysis,

Table 11-3. Existing Building LOADCAL Floor Summary

MAXIMUM HOURLY VALUES

MON	MAX HR COOL	MAX HR HEAT
JAN	112,845	-19,038
FEB	114,438	-15,449
MAR	107,624	-10,868
APR	105,940	-4,739
MAY	123,642	0
JUN	133,735	0
JUL	138,024	0
AUG	122,146	0
SEP	120,032	0
OCT	124,459	-4,036
NOV	118,523	-12,907
DEC	112,524	-18,725
ANN	138,024	-19,038

TOTALS

MON	COOL LOAD	HEAT LOAD	TOTAL LOAD
JAN	80,336,345	-10,140,689	90,477,034
FEB	78,392,888	-6,739,516	85,132,404
MAR	94,761,916	-4,340,496	99,102,412
APR	101,792,160	-737,040	102,529,200
MAY	117,195,531	0	117,195,531
JUN	124,452,780	0	124,452,780
JUL	142,220,374	0	142,220,374
AUG	133,539,599	0	133,539,599
SEP	120,979,860	0	120,979,860
OCT	108,031,683	-872,619	108,904,302
NOV	87,417,210	-5,034,060	92,451,270
DEC	80,701,494	-9,707,991	90,409,485
ANN	1,269,821,840	-37,572,411	1,307,394,251

ITEM SUMMARY

MON	LIGHTING	PEOP LAT	PEOP SEN	VENT LAT	VENT SEN	ROOF	GLASS CON	GLASS SHG	WALL
JAN	71,741,471	7,502,000	7,492,452	966,580	3,037,659	0	43,328,080	37,106,628	6,314,576
FEB	64,798,748	6,776,000	6,767,375	781,200	2,379,132	0	35,384,384	36,490,552	4,634,588
MAR	71,741,471	7,502,000	7,492,452	356,500	1,964,749	0	32,039,492	41,635,666	3,589,428
APR	69,427,230	7,260,000	7,250,760	1,427,400	763,590	0	18,443,760	38,875,290	1,232,550
MAY	71,741,471	7,502,000	7,492,452	151,900	444,664	0	9,397,216	37,992,549	1,004,365
JUN	69,427,230	7,260,000	7,250,760	1,917,900	788,910	0	6,372,720	35,669,520	2,357,600
JUL	71,741,471	7,502,000	7,492,452	4,218,480	1,561,036	0	8,520,560	36,643,066	4,157,565
AUG	71,741,471	7,502,000	7,492,452	2,697,280	1,007,159	0	6,487,680	38,047,633	2,538,980
SEP	69,427,230	7,260,000	7,250,760	1,131,000	521,310	0	7,251,840	38,380,470	1,651,710
OCT	71,741,471	7,502,000	7,492,452	1,219,230	655,774	0	17,377,360	38,277,219	1,701,652
NOV	69,427,230	7,260,000	7,250,760	395,800	1,834,980	0	30,529,280	35,195,150	3,589,970
DEC	71,741,471	7,502,000	7,492,452	915,740	2,900,422	0	41,899,600	36,121,138	6,147,796
ANN	844,697,965	88,330,000	88,217,580	16,075,010	17,859,385	0	257,537,972	450,635,321	39,650,764

the typical office floor was divided into five zones, with each perimeter zone having one orientation. In this analysis, the perimeter zones are further subdivided into bays, or rooms. This is done in order to effectively determine the available day lighting.

Each bay was analyzed to determine dimensions, ceiling heights and surface reflectance (ceiling, wall and floor) values. Specific window conditions including orientation, area, transmittance value and blind usage were established. Direct sunlight could not enter the space due to the positive stop provided with the horizontal blinds. Typical adjustment angle was observed to be approximately 45 degrees.

Actual Atlanta meteorological data were used in this analysis. Solar altitude angles and azimuths were calculated for specific times and then window-to-sun orientations were established.

Illumination levels, at the windows of each bay were determined for various times of the day, under clear and cloudy sky conditions. Coefficients of utilization were established based upon the specific bay and window conditions. From these factors, daylighting levels were calculated on three work surfaces, located within each bay. Daylighting levels were calculated for the following times: 8:00 a.m., 10:00, 12 noon, 2:00 and 4:00 p.m., for March 21st, June 21st and December 21st. (Note: illumination levels for September 21st equals the levels for March 21st.)

A computer spreadsheet was developed to calculate the daylighting levels described above. The appendix contains a sample of the spreadsheet, bay number one at March 21st under clear and overcast sky conditions, used to determine the day lighting levels. The procedure is similar for the other times and for the other three bays. Following are graphs which describe the daylighting levels for the times and sky conditions noted on each graph.

The daylighting levels, which vary depending upon bay orientation, time of day and sky condition, will be the basis for evaluating any modification to the existing building. The strategy is to utilize the natural day lighting available to reduce artificial lighting levels, thereby improving the thermal performance of the structure, which will reduce the operating costs.

REVISED BUILDING SCHEME DESCRIPTION

As stated earlier, the goal of this investigation is to analyze the implications of using daylighting on a building's thermal load. In an effort to effectively analyze this issue, without having to consider the possible interactive effects of multiple alterations, the basic building design was not significantly altered. Changes to the overall design, such as modifications to floor height, window configuration, overhang or vertical fin depths were not considered. The addition of reflectors or sun screens was not considered either. Building usage patterns nor operational procedures could not be altered. With these constraints established, as the basis for defining alternative solutions, the only remaining element for consideration is the window glazing.

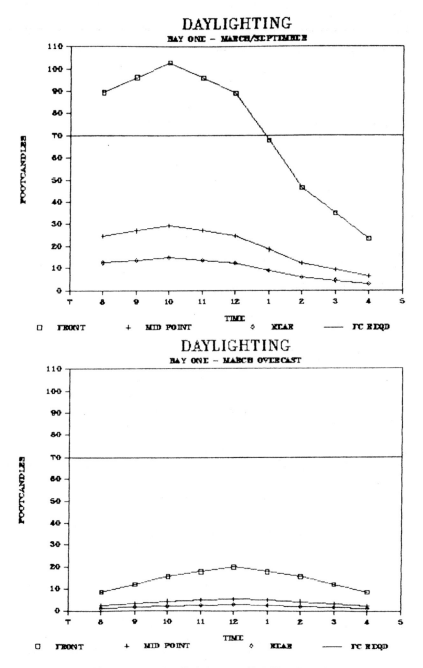

Figure 11-1. Daylighting Availability Graph

The existing glazing is a bronze-tinted, double insulating glass, "Solarban 550-60" as manufactured by PPG. It has a shading coefficient of .46, a visible light transmittance of 32 percent and a U-value (winter) of .50. It was a good glass selection, considering the types available when the facility was constructed.

Now, however, glazings with greatly improved performance characteristics are available. Low-emissivity glass and specialty coated glazings are available. One manufacturer has developed a product line called "Low-E" glass, after this new technology. The price range of these new glazings is comparable to existing insulating glass.

In selecting a new glazing type for this building, the following factors were evaluated for the types available today: the shading coefficient, the visible light transmittance and the U-value of the glazing.

The new glass should have the lowest shading coefficient factor available and the highest visible light transmittance. This will provide the least amount of solar heat-gain in relation to the amount of daylighting entering the space. By having a low shading coefficient and high light-transmittance, day lighting is provided while the heat associated with the light is filtered out.

The U-value, of course, needs to be as low as possible to cut down on the conductive heat losses/gains to/from the environment. Eliminating these adverse effects will make the space 'feel' more comfortable for cold drafts in winter and radiated heat in summer will be reduced.

The methodology for selecting a new glazing type, at the outset, was to select a glazing with a shading coefficient comparable to the existing and a visible light transmittance factor as high as possible. While this approach was sound from the standpoint that existing artificial lighting levels could be greatly reduced, the amount of natural daylighting provided was excessive and, thus unacceptable. The lighting levels generated were so great that glare and over-illumination would have made the space impossible to work in, especially at the window locations.

The opposite approach was then investigated. By selecting a glazing with a transmittance factor comparable to the existing and the lowest available shading coefficient, the results were more acceptable. While the availability of natural daylighting did not change significantly (it was actually decreased by six percent from the levels available in the existing design), the solar gain associated with the daylight was decreased dramatically.

The glazing selected is manufactured by the Interpane Group, and comes from their IPASOL solar reflective, insulating glass line. The glass is the IPASOL Gold 30/17. It is a double-insulating glazing with a gold-tinted low-emissivity coating applied to the number three surface. The visible light transmittance is 30 percent, lower than the existing glass by six percent. The shading coefficient factor is .19, which is 59 percent lower than the existing. The U-value (winter) is also much lower at .23 compared to .50 for the existing.

This one modification has the potential of providing considerable energy savings, if only by improving the thermal performance of the glazing portion of the building. However, when accounting for the artificial lighting level reductions possible by incorporating the available daylighting, an even greater thermal load reduction is probable.

Table 11-4. Glass Comparison Data

Existing Glass:	
U-value (summer)	= .59
U-value (winter)	= .50
Shading Coefficient	= .46
Visible Light Transmission	= 32%
Glazing Selection One:	
U-value (summer)	= .33
U-value (winter)	= .32
Shading Coefficient	= .42
Visible Light Transmission	= 61%
Glazing Selection Two:	
U-value (summer)	= .23
U-value (winter)	= .23
Shading Coefficient	= .19
Visible Light Transmission	= 30%

* = Glazing selected for Building Modification.

REVISED DAYLIGHTING ANALYSIS

Daylighting data were developed for the revised building scheme. This procedure involved factoring the existing building daylighting values by the six percent reduction of the transmittance factor.

Once these values were determined, an average illuminance availability was calculated for each bay. This was done for the times, months and sky conditions described in the earlier daylighting analysis. Yearly illumination levels were then established for each bay. These yearly values were factored by a 'usage factor' based on day lighting availability in relation to lighting requirements, in terms of hours per day. Then, these yearly averages were compared against the lighting level requirement of 70 footcandles.

From these calculations, reductions in the existing artificial lighting level, expressed as percentages, were determined. A 15 percent reduction in the artificial lighting level is possible within bays one and two. In bays three and four. a nine percent reduction is possible.

It must be noted that, while incorporating the available daylighting will allow for reduced artificial lighting levels, they cannot be permanently reduced. A control device, or devices, will be required to increment the amount of artificial lighting reduction, based upon the contribution of daylighting at any given time. This is due to the ever changing nature of daylighting.

REVISED LOADCAL ANALYSIS

Data concerning the modified building, the new U-value, the new shading coefficient and the reduced artificial lighting levels, were entered into the LOADCAL program for analysis. All other building usage and design factors remained as described in the original analysis. The results, in terms of reduced thermal load, are impressive.

The results are described in the following chapter. The LOADCAL analysis data will be contained in the appendix. It will be presented in the same format as provided for the original building analysis.

RESULTS

By modifying the existing building with the new glazing and incorporating the available daylighting, to reduce the artificial lighting level significant reductions in the estimated energy consumption are realized.

The much-improved U-value (winter) of .25 of the new glass yielded a reduction of 53.99 percent in the glazing conductive heat transfer. While being significant in terms of the amount of actual heat loss/gain, it will also have the effect of increasing the comfort level of the space. This is accomplished by cutting down on cold down-drafts near the windows in the winter and conductive heat gain in the summer.

The lower shading coefficient provided a 58.95 percent reduction in the solar heat gain to the space. This is a considerable reduction, especially when one realizes that the amount of natural daylighting entering the building was reduced only six percent. (The visible light transmittance is 30 percent for the new glass, while the existing glass was 32 percent.)

The reduced thermal load afforded by lower artificial lighting levels, is 10.51 percent. While this may not seem significant, when compared to the reductions above, it is a tremendous amount of energy considering the

Table 11-5. Revised Building LOADCAL Floor Summary Data

MAXIMUM HOURLY VALUES

MON	MAX HR COOL	MAX HR HEAT
JAN	61,931	-7,929
FEB	63,180	-5,719
MAR	61,823	-3,361
APR	71,363	-80
MAY	80,811	0
JUN	87,352	0
JUL	91,775	0
AUG	83,854	0
SEP	74,001	0
OCT	70,930	0
NOV	66,033	-4,578
DEC	61,964	-7,753
ANN	91,775	-7,929

TOTALS

MON	COOL LOAD	HEAT LOAD	TOTAL LOAD
JAN	67,134,003	-2,787,923	69,921,926
FEB	64,161,188	-1,583,624	65,744,812
MAR	76,496,995	-698,647	77,195,642
APR	80,936,670	-2,400	80,939,070
MAY	92,085,097	0	92,085,097
JUN	96,374,370	0	96,374,370
JUL	108,405,822	0	108,405,822
AUG	102,503,298	0	102,503,298
SEP	93,611,730	0	93,611,730
OCT	86,604,824	0	86,604,824
NOV	71,981,040	-942,900	72,923,940
DEC	67,564,717	-2,625,297	70,190,014
ANN	1,007,859,754	-8,640,791	1,016,500,545

ITEM SUMMARY

MON	LIGHTING	PEOP LAT	PEOP SEN	VENT LAT	VENT SEN	ROOF	GLASS CON	GLASS SHG	WALL
JAN	64,203,697	7,502,000	7,641,190	966,580	3,037,659	0	19,931,264	15,249,272	6,314,576
FEB	57,990,436	6,776,000	6,901,720	781,800	2,373,132	0	16,278,192	14,982,520	4,634,588
MAR	64,203,697	7,502,000	7,641,190	356,500	1,964,749	0	14,739,570	17,101,706	3,589,428
APR	62,133,610	7,260,000	7,394,700	1,427,400	763,590	0	8,486,020	15,942,380	1,232,550
MAY	64,203,697	7,502,000	7,641,190	151,900	444,664	0	4,323,384	15,583,111	1,004,369
JUN	62,133,610	7,260,000	7,394,700	1,917,900	788,910	0	2,931,480	14,612,070	2,357,600
JUL	64,203,697	7,502,000	7,641,190	4,218,480	1,551,035	0	4,102,574	15,116,576	4,157,555
AUG	64,203,697	7,502,000	7,641,190	2,893,280	1,097,159	0	2,964,246	15,558,146	2,838,380
SEP	62,133,610	7,260,000	7,394,700	1,131,000	521,310	0	3,336,240	15,795,900	1,651,710
OCT	64,203,697	7,502,000	7,641,190	1,215,030	655,774	0	7,994,156	15,728,811	1,701,652
NOV	62,133,610	7,260,000	7,394,700	855,800	1,834,950	0	14,085,860	14,461,440	3,969,970
DEC	64,203,697	7,502,000	7,641,150	915,740	2,900,422	0	13,074,560	14,831,051	6,147,798
ANN	755,946,755	88,330,000	89,968,850	16,075,010	17,859,385	0	118,473,946	185,005,385	39,660,784

overall lighting level of 70 footcandles is still provided at each work surface. And when one realizes the amount of electrical consumption saved, by the artificial lighting reduction, the actual energy savings are significant.

Another significant reduction that this modification affords is a 33.57 percent reduction in the peak-hourly cooling load of the space. While this will save considerably in the overall energy consumption, it will yield great economic savings on the amount of peak-electrical demand charges applied to the electrical consumption. It could also provide considerable construction cost savings since all mechanical equipment must be sized to handle the maximum hourly cooling load. Should the owner decide to install the new glazing as a retrofit program and reduce the artificial lighting levels within the building, additional savings could be realized by rebalancing the mechanical system. The greatest potential savings would, however, result in applying this analysis to the design of a totally new facility.

Finally, the total thermal load reduction afforded by these modifications is 22.25 percent. This is quite a significant reduction when one considers the simple modifications that were suggested.

Discussion of the Results:

The amount of thermal load reductions and economic savings provided are significant by any standard. This is especially true considering that the suggested modifications, to the existing building, are very minor compared to methodologies typically employed to reduce energy consumption levels of a structure. As noted earlier, the cost of the new glazing is comparable with other insulating glazings, and when factored for inflation, not significantly more expensive than the existing glazing in the building.

The economic justification, for making this modification to the facility, has been proven with the reduced thermal loading, lower electrical consumption rates for artificial lighting and cooling loads, reduced peak-load electrical surcharges and reduced mechanical plant size requirements.

CONCLUSIONS

In conclusion, this analysis proves that great reductions in a structure's energy consumption can be achieved with little change in its appearance or construction. It is possible to achieve the desired results of a more energy efficient building without significantly increasing the initial construction costs. The results of this analysis also indicate any increase in the initial construction costs can be recovered in the reduced operating costs of the facility.

Furthermore, this study indicates that initial choices and decisions in the design phase directly affect the overall energy performance of a structure. It is imperative, therefore, that architects and engineers analyze design choices and decisions for their possible effect upon the structure's thermal performance. It is also important that owners and developers realize that initial construction costs are not always the most important factor in a building's design. The performance characteristics and operating costs should be addressed. As energy prices begin their upward spiral, as predicted, the service of providing energy efficient structures with lower operating costs will become even more important.

RECOMMENDATIONS

The analysis procedure outlined in this chapter, while somewhat lengthy, would not be too burdensome to incorporate into the initial design phase of a structure. The author believes that it can serve as an effective analysis tool to evaluate proposed designs and / or modifications to facilities such as the one described herein. It is recommended that interested persons utilize this procedure, as another design determinant, in the overall design process.

It is also recommended that further research into this aspect of the design process be undertaken to streamline the procedure outlined within. This would be a valuable service to all parties interested in providing energy efficient structures and facilities. Finally, it is hoped that further research into technologies and methodologies be conducted that will allow even more energy efficient structures to be designed and constructed.

References
1.   American Society of Heating, Refrigerating and Air-Conditioning Engineers, Inc.; *Cooling and Heating Load Calculation Manual*; New York; ASHRAE; 1979; Second Printing.
2.   American Society of Heating, Refrigerating and Air-Conditioning Engineers, Inc.; *Handbook—1977 Fundamentals*; New York; ASHRAE; 1979; Second Printing.
3.   American Society of Heating, Refrigerating and Air-Conditioning Engineers, Inc.; *Handbook—1978 Applications*; New York; ASHRAE; 1979; Second Printing.

4. American Society of Heating Refrigerating and Air-Conditioning Engineers, Inc.; *Handbook—1979 Equipment*; New York; ASHRAE; 1979; Second Printing.

5. American Society of Heating, Refrigerating and Air-Conditioning Engineers, Inc.; *Handbook—1977 Systems*; New York; ASHRAE; 1979; Second Printing.

6. Libbey-Owens-Ford Company; *How to Predict Interior Daylight Illumination*; Ohio; LOF; 1976.

7. Jarrell, R. Perry; "Special Problems Research: A comparison of Computer Estimation Programs"; Class Paper; 1985.

8. Aeck Associates, Architects; "Contract Documents for the Construction of Twin Office Towers, Project No. GBA-39, Georgia State Financing & Investment Commission"; Atlanta, Georgia; 1977.

9. Georgia Building Authority (GBA) "Construction, Energy Consumption and Occupancy Records for the Twin Office Towers"; Atlanta, Georgia; 1984.

10. Akridge, Professor Max; et al.; "Class Notes for the Energy in Architecture Series"; Georgia Institute of Technology; 1984.

Chapter 12

Windows and Daylighting*

S.E. Selkowitz, D. Arasteh, D.L. Dibartolomeo,
A.J. Hunt, R.L. Johnson, H. Keller, J.J. Kim,
J.H. Klems, C.M. Lampert, K. Lottus,
K. Papamichael, M.D. Rubin,
M. Spitzglas, R. Sullivan,
P. Tewari, and C.M Wilde

Approximately 20% of annual energy consumption in the United States is for space conditioning of residential and commercial buildings. About 25% of this amount is required to offset heat loss and gain through windows. In other words, 5% of U.S. energy consumption-the equivalent of 1.7 million barrels of oil per day-is tied to the performance of windows. Fenestration performance also directly affects peak electrical demand in buildings, sizing of the heating, ventilating, and airconditioning (HVAC) system, and the thermal and visual comfort of building occupants.

The aim of the Windows and Daylighting Group of Lawrence Berkeley Laboratory (LBL) is to develop a sound technical base for predicting the net thermal and daylighting performance of windows and skylights. The group's work will help generate guidelines for design and retrofit strategies in residential and commercial buildings and will help develop new high-performance materials and designs.

One of Lawrence Berkley Laboratory's (LBL) program strengths is its breadth and depth: at one extreme, LBL can examine energy-related aspects of windows at the atomic and molecular level in its materials science studies, and at the other extreme, LBL can perform field tests and in-situ experiments in large buildings. They have developed, validated, and use a unique, powerful set of computational tools and experimental facilities. Their scientists, engineers, and architects work in collaboration

*This work was supported by the Assistant Secretary for Conservation and Renewable Energy, Office of Buildingsand Community Systems, Building Systems Division and Office of Solar Heat Technologies, Solar Buildings Division of the U.S. Department of Energy under Contract No. DE-AC03-76SF00098.

with researchers in industry and academia.

To be useful, the technical data developed by LBL's program must be communicated to design professionals, industry, and other public and private interest groups.

LBL research is organized into three major areas:

- Optical Materials and Advanced Concepts
- Fenestration Performance
  - Thermal analysis
  - Daylighting analysis
  - Field measurement facility
  - Building monitoring
- Building Applications and Design Tools
  - Nonresidential studies
  - Residential studies
  - Occupant impacts
  - Design tools

DAYLIGHTING STUDIES

Providing daylight to building interiors is one of fenestration's most important functions, both from an energy perspective and from an occupant's point of view. LBL conducts a wide range of activities to establish the facilities, tools and data to address these perspectives.

Daylighting Optics of Complex Glazing and Shading Systems

A quantitative understanding of the solar-optical properties of fenestration systems is essential to accurately calculate daylight illuminance levels, glare potential, solar heat gain, and thermal comfort. Existing models are adequate for determining the properties of simple, but not complex, fenestration systems.

LBL wants to develop a methodology to analyze the luminous and solar heat gain performance of any complex fenestration system. They determine daylight performance by treating fenestration systems as light sources of varying candlepower distribution that can be calculated from bidirectional solaroptical properties. This information can be used in a daylighting model to predict interior illuminance distribution. They derive solar heat gain by processing the directional-hemispherical solar-optical properties to get the total transmitted and absorbed radiation. For these purposes, they have developed two new experimental facilities, a scanning photometer radiometer and an integrating sphere.

The scanning photometer/radiometer is used to determine the bidirectional transmittance and reflectance of fenestration components and systems, i.e., the fraction of the incoming radiation from any single direction that is transmitted or reflected toward any single outgoing direction.

LBL uses the integrating sphere to determine the directional hemispherical transmittance of fenestration components and systems, i.e., the fraction of the incoming radiation from any single direction that is transmitted toward all outgoing directions. These two facilities are part of a Daylighting Laboratory that includes a roof-top station for measuring daylight availability, outdoor scale-model test facilities, and a 24-ft-diameter sky simulator for making scale model building measurements under controlled conditions.

LBL developed a new computer program. TRA (Transmittance Reflectance Absorptance), which calculates the bidirectional solar-optical properties of complex fenestration systems given the bidirectional solar-optical properties of their component layers. Additional software was partly implemented for determining the hemispherical-hemispherical and the hemisphericaldirectional solar-optical properties of fenestration systems for the CIE overcast and clear sky luminance distributions, given their directional-hemispherical and bidirectional solar-optical properties. These data will serve as input to the SUPERLITE and the WINDOW computer programs for determing daylight and thermal performance, respectively.

LBL will continue developing and documenting the methodology, concentrating on new detectors and more versatile computerized data collection system for the scanning radiometer and the required software for the manipulation of bidirectional solar-optical properties. Their goal is to entirely automate the procedures for data collection and manipulation to generate a data base for a large number of fenestration systems. This data base will be input to energy analysis programs such as DOE-2.

LBL also intends to intervalidate their experimental facilities and analytical routines.

The Sky Simulator and Daylight Photometric Laboratory

A 24-ft-diameter hemispherical sky simulator (Figure 12-1) was designed and built on the University of California's Berkeley campus in 1979. In operation since 1980, it can simulate uniform, overcast, and clear-sky luminance distributions. Sky luminance distributions are reproduced on the underside of the hemisphere; light levels are then measured in a scale-model building at the center of the simulator. From these measurements LBL can accurately and reproducibly predict day lighting illuminance patterns in real buildings and thereby facilitate the design of energy-efficient buildings. The facility is used for research, for educational purposes, and on a limited basis by architects working on innovative daylighting designs.

Daylighting Analysis
In previous years LBLdeveloped several simplified daylighting design tools (such as the QUICK LITE program), that are now widely used in the architectural and lighting design communities. Last year they

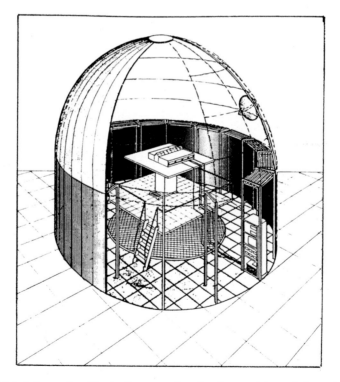

Figure 12-1. Schematic of 24-ft-diameter sky simulator with model on platform. (XBL 8412-5328)

expanded the range of modeling capabilities and improved computational accuracy in SUPERLITE, their advanced program. A collaborative effort to develop a new lighting/daylighting model was also initiated.

LBL continues to test and evaluate SUPERLITE, concentrating on its ability to model complex shading systems. Their approach is to define the daylight transmittance properties of the window and shading system as a candlepower distribution function that varies with sky conditions and/ or the indicence angle of sunlight. LBL initially used theoretical distributions that could be compared to results generated by other computational techniques. This comparison showed good agreement.

LBL continues to test a version of SUPERLITE with an electric lighting modeling capability. This new capability will allow LBL to study the combined effects of daylight and electric light in a room.

LBL also worked jointly with the ABACUS group, University of Strathclyde, Scotland, to develop an improved lighting/daylighting model for their building simulation program, ESP. This model, still under development, calculates spectral data so that illuminance results can be accurately reproduced on a color computer monitor. It can also model specular surfaces. Linked to a graphics package, which displays 3-D room views, it should provide a significant advance in LBL's ability to accurately represent illuminated interior spaces.

LBL will continue testing and validation of SUPERLITE's modeling of shading devices, using candlepower data sets measured by the luminance scanner. Collaboration on development of the new ESP lighting model will continue.

Coefficient-of-Utilization
Model for Energy Simulation Models

Building energy analysis computer models must be able to predict the daylighting performance of complex design strategies commonly used by innovative architects. The models should either internally calculate the daylight illumination or be supplied with data precalculated by other illumination models or measured in scale models. The first approach, internal calculation of daylight illumination, is generally impractical for complex designs because of the significant computational cost and complexity required to obtain reasonably accurate answers. LBL is therefore developing a coefficient-of-utilization (CU) model that will be compatible with an hour-by-hour simulation model but still retain the flexibility and accuracy of more complex computational models.

LBL's approach is to derive the CU model from regression analysis of a parametric series of day lighting analyses using SUPERLITE. They previously modified SUPERLITE to generate these parametric series and also modified the output to report the indoor illuminance level due to each external light source—sun, sky, or ground.

They used a statistical computer software package to generate test regression equations for the sun, sky, and ground with a limited number of generalized variables (e.g., location of window, window dimensions). Initial results suggest that good fits can be obtained with relatively simple regression expressions. However, more extensive analysis is required to extend this to a wide range of room and fenestration conditions.

LBL will expand the effort to develop new CU models that can handle greater variation in room geometry and surface reflectance. After they generate all equations, the equations will be thoroughly tested and validated. The final model will be incorporated into future versions of DOE-2 and may also be a stand-alone illuminance model.

Daylight Availability Studies

Accurate daylight availability models are necessary for many design and energy analysis simulations. In 1978 LBL began an availability measurement project, as data were lacking for most U.S. locations. However, a widely accepted generalized model of availability in the U.S. has yet to be developed.

They previously published three papers analyzing daylight availability data for San Francisco. Analysis focused on the relationship of measured illuminance and irradiance to atmospheric parameters such as turbidity. A new functional relationship was developed to determine an illuminance turbidity for visible radiation analogous to the conventional turbidity terms used with solar radiation. LBL also developed new functional relationships for zenith luminance as a function of turbidity and found that their clear-sky luminance distribution data agree well with data from currently accepted European models. These results were published in the Technical Proceedings of the International Daylighting Conference.

In 1986 LBL's focus shifted to developing a better understanding of the nature of partly cloudy skies. Using a trailermounted sky luminance mapper developed by the Pacific Northwest Laboratories (PNL) they measured sky luminance profiles every few minutes throughout the day on the roof of the Space Sciences Laboratory in the Berkeley hills. Analysis of the data is being completed at Florida Solar Energy Center (FSEC).

LBL will continue to analyze their existing data base and complete a study of the luminous efficacy of daylight and sunlight. Data collection with the sky luminance scanner will be completed. They will collaborate with FSEC researchers on analysis of the availability data but will reduce the overall activity in this area at LBL.

BUILDING APPLICATIONS AND DESIGN TOOLS

Research to develop new glazing materials and to better understand fenestration performance will provide real energy savings only if the technology is effectively applied in buildings. Using the technology requires that LBL has detailed understanding of how a wide range of fenestration systems can be optimally used in different building types and climates, and that they pass this understanding, through design tools, to building design professionals.

SIMULATION STUDIES: NONRESIDENTIAL BUILDINGS

While most building energy simulation studies have focused on minimizing total energy consumption, other issues are equally important. Peak electrical demand affects both user costs and the utilities' required generating capacity. A complete study of the cost effectiveness of fenestration systems, particularly those incorporating daylighting strategies, must include their impact on peak electrical demand as well as on energy savings. In addition, issues such as comfort and convenience affect the user acceptance and, consequently, effectiveness of these systems. LBL's studies explore the interactions of these Issues.

During the past several years the effects of a wide range of glazing properties, window sizes, lighting loads, orientations, and climates on the energy performance of a prototypical office building have been simulated with DOE-2.1B. Initially LBL examined the impact of fenestration properties, including the effects of daylighting strategies, on office building energy performance and peak electrical demand. Lighting energy savings resulting from daylighting were examined for a range of fenestration properties and lighting control systems. Annual energy consumption of an office module was found to be sensitive to variations in U-value, shading coefficient, and visible transmittance, as well as glazing area, orientation, climate, and operating strategy. Sample results from our simulation studies are shown in Figure 12-2.

LBL concludes that, in almost all instances, it is possible to find a fenestration design strategy that outperforms a solid insulating wall or roof and that daylighting is almost always an essential component of energy savings. If the installed electric lighting power density is high, the energy savings potential is large. More efficient electric lighting systems reduce day lighting benefits.

The net benefits of fenestration are highly dependent on the tradeoffs between daylighting savings and cooling loads resulting from solar gains. Visible transmittance properties, improved shading design, and window

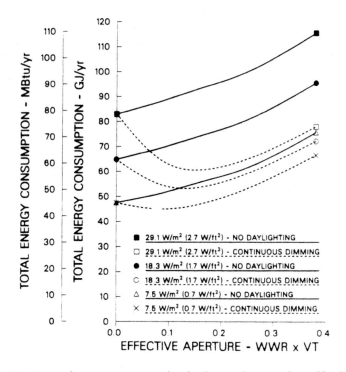

Figure 12-2. Annual energy consumption in the south zone of an office building in Madison, Wisconsin, as a function of window area. "Continuous dimm ing" indicates controlled reductions in electric lighting in response to daylight; "no daylighting" indicates no dimming. (XBL 876-2796)

management will thus assume increasing importance for maximizing energy benefits from daylight. LBL's studies have demonstrated that the common assumption that daylighting is a "cooler" source of light than electric lighting is not necessarily true. LBL has developed and is refining a methodology for comparing cooling loads imposed by daylight (or electric light) through the use of an index derived as a fraction of three parameters:

(1) The relative T vis and SC of the glazing/shading system;
(2) The distribution of daylight within the space;
(3) The time-dependent absolute transmitted solar intensity.

LBL continued their general studies on the peak-shaving potential of daylighting with results that show-despite solar gains-daylighting can significantly reduce peak electrical demand during summer months. The critical tradeoffs-between electric lighting reductions from daylighting and cooling load increases from solar gain-help determine the combination of window properties that minimize building peak loads. A breakdown of this load for a sample office building at the hour of peak demand is shown in Figure 12-3 for both daylighted and nondaylighted

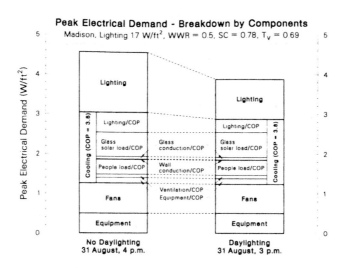

Figure 12-3. A comparison of peak electric demand, by component, for an office building in Madison, WI, with and without daylight utilization. (XCG 855-223)

cases. In Figure 12-4, annual peak electrical demand is shown for LBL's prototypical building in Lake Charles, LA. The benefits of daylight use are clearly illustrated.

Simulations of advanced glazing materials having active response functions, represented in the lower curve in Figure 12-5, show large potential savings.

Skylights can also provide significant energy and cost benefits. Daylighting benefits are maximized with relatively small ratios of sky-

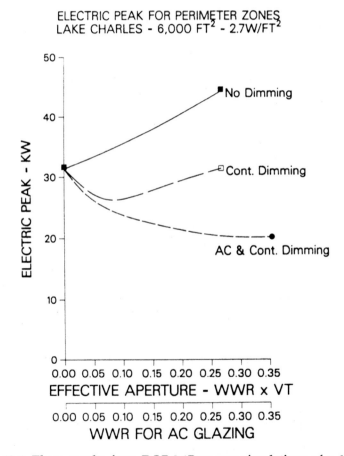

ELECTRIC PEAK FOR PERIMETER ZONES
LAKE CHARLES - 6,000 FT^2 - 2.7W/FT^2

EFFECTIVE APERTURE - WWR x VT

WWR FOR AC GLAZING

Figure 12-4. These results from DOE-2.1B energy simulations of a 6000 ft^2 office building in Lake Charles, LA, show the effect of day lighting on peak electrical demand as a function of effective aperture. Electric lighting density is 2.7 W/ft^2. (XBL 876-2802)

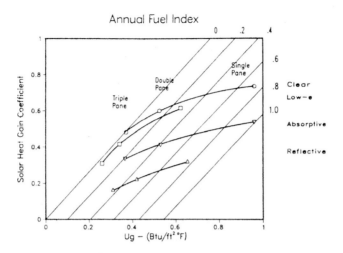

Figure 12-5. Annual fuel index as a function of glazing parameters. (XBL 876-2804)

light to roof areas (0.01-0.04). Because skylights provide more uniform daylight distribution, the cooling load impact of daylighting is less than with vertical fenestration. As effective aperture is increased beyond the optimum, cooling loads in most climates rise to adversely affect net annual energy performance.

A large number of DOE-2.1 B runs for window and skylight studies have provided enough data for multiple regression techniques to develop analytical expressions of energy requirements as functions of glazing parameters; from these LBL may be able to develop a generalized expression to accommodate climate variables. The simple expressions correlate well with DOE-2 results and may become a design tool to assess energy and cost trade-offs among fenestration options. LBL has incorporated the regression equations developed in their skylight studies in a skylight design handbook sponsored by AIA, AAMA, and NFC, and this procedure is the basis for the LRI Fenestration performance Indices described in the next section.

Simulation studies to date have examined the energy impacts of many fenestration parameters for conventional designs and have begun to explore the potentials of new optical materials. New studies will examine the performance of shading devices for which adequate solar optical data do not exist. Optical properties of shading devices will be measured in LBL's laboratory. They will also continue to look at variations in window shade energy performance and management strategies, issues of daylight luminous efficacy, advanced glazing materials, the effects of fenestration performance on HVAC, and the effects of various BY AC options on fenestration performance. The costs of fenestration design and daylighting as influenced by peak electrical demand, annual energy use, and chiller size will be examined.

LRI FENESTRATION PERFORMANCE INDICES

Building designers, utility auditors, and others are constantly required to compare and evaluate the performance of alternative fenestration systems in buildings. To properly address the energy-related impacts of fenestration, one must be able to quantify energy performances for all systems in a systematic and reproducible way. The objective of this study for the Lighting Research Institute Center (LRI) is to develop numerical indicators of fenestration system performance on the basis of annual energy consumption, peak electrical demand, illumination performance, and themlal and visual comfort. These indicators are to be used as guides in evaluating and selecting alternative fenestration products and systems for use in various building types and climates. In Phase 1 LBL is developing the basic methodology for determining the performance indicators; Phase 2 will support the measurement and analysis required to construct a microcomputer design tool that embodies the project's results.

Using thermal and visual comfort indices in an energy design tool is a major objective of this project. A methodology was developed that defines the relationship between fenestration characteristics, direct solar radiation, and thermal and visual comfort so that annual comfort indices can be calculated. The thermal comfort index is related to mean radiant temperature within the space. Visual comfort is based on a glare index calculated by DOE-2.

Numerous DOE-2 runs have been completed for a prototype office building in Madison, WI, and Lake Charles, LA. Simplified design charts were developed from the multiple regression coefficients obtained from these runs. Figure 12-5 shows the variation of an annual fuel index as a function of solar heat gain coefficient and window U-value. Four glazing types are shown as well as the variation in number of panes of glass. Similar figures have also been generated for indices of annual electric energy and peak electric load.

Most complex fenestration products and systems cannot be readily characterized using conventional analysis techniques, so LBL is determining the solar-optical properties and daylight transmittance/distribution functions of these systems experimentally, using the integrating sphere and the luminance/radiance scanner described previously.

To complete Phase I, LBL will develop a methodology for incorporating the proposed performance indices into an overall fenestration system figure-of-merit. A weighting function will be provided so that users can assign specific relative weights to the indices they deem important. If, for example, a designer wishes to maintain comfort without mechanical cooling, minimizing the cooling requirement might be a priority task. Thus, the cooling and thermal comfort index might be heavily weighted and an appropriate fenestration option selected accordingly.

A workshop for industry representatives will be conducted to present the methodology and performance indices developed in Phase I. Evaluation results from the workshop will help LBL plan the development of the microcomputer design tool that is the key product of Phase 2, which they hope to initiate by the end of the year.

SIMULATION STUDIES:
RESIDENTIAL BUILDINGS

LBL's studies have focused on techniques to simplify accurately the very complicated heat transfer processes that occur between the components of a building. They have previously shown the feasibility of isolating window systems from other building components such as envelope insulation levels, infiltration, and internal heat gains in determining building energy performance.

The regression expressions developed in past years to predict residential energy use as a function of fenestration parameters were expanded to include effects from the use of night insulation, shade management, and overhangs. For a given climate, orientation, and window size, LBL developed graphic plots that allow one to quickly evaluate seasonal performance differences in alternative existing or hypothetical fenestration systems.

In addition, LBL developed a procedure to directly compare the thermal loads resulting from different building prototype configurations independent of geographic location, to ultimately help LBL develop a design tool with broad climatic applicability.

LBL used these techniques to describe the performance of low-E windows in residences in hot and cold climates. This study analyzed the energy and cost implications of conventional double- and triple-pane windows and newer designs in which substrate, type, and location of low-E coating, and gas fill, were varied. The analysis showed the potential for substantial savings but suggested that both heating and cooling energy should be examined when LBL evaluates the performance of different fenestration systems. The study also showed the importance of considering window frame effects for the lowconductance glazing units.

Most analysis procedures describe the performance of specific alternatives and one must analyze many cases to arrive at a solution that apperas to optimize energy use. LBL developed a mathematical technique to directly calculate the window size or properties that minimize energy use or cost for a given climate and orientation. It will become part of the design tools developed next year.

LBL's ultimate objective is to develop handbooks, charts, nomographs, and computer software that will help builders, developers, and suppliers assess different window strategies. The immediate objective in 1987 was to create a prototype residential fenestration design tool that would be used to stimulate building industry interest in collaborative development of a comprehensive tool.

DESIGN TOOLS
AND TECHNOLOGY TRANSFER

To influence energy consumption trends in the United States, it is critical for LBL to package and transfer their results to other researchers and professionals. The needs and motivations of this group, including designers, engineers, building owners, manufacturers, and utilities, vary widely, so LBL uses a variety of media to reach each audience. Their activities have included developing improved daylight analysis and design tools, design assistance studies, occupant response studies, workshops, conferences, assessment reports, handbooks, and meetings with industrial and design firms and utilities. Other efforts have been designed to communicate results of their work widely to other research and development groups, educational institutions in the U.S. and abroad, and professional and industrial societies.

Design Tools

The Windows and Daylighting Program and subcontractors develop and distribute daylighting design tools to industry and educational institutions. Private-sector software firms continue to introduce new design tools for daylighting but it is difficult for a potential user to evaluate and compare them. Thus there is: (1) for a comparative matrix identifying existing tools, and (2) for photometric data and evaluation procedures to compare these tools.

Development of a general procedure for comparing design tools predictions to a photometric data base derived from sky simulator measurements and SUPERLITE simulations continued. The Daylighting Design Tool Survey was published, describing the capabilities of more than 30 tools including nomographs, protractors/tables, and micro, mini, and mainframe computer tools. The design tools tested and disseminated in 1986 are described below:

Clear Sky Protractors. Developed with Harvey Bryan at the Massachusetts Institute of Technology (MIT), the protractors consist of a series of transparent overlays that are placed over architectural plans and elevations to determine a window's daylight contribution at any point in a room.

SUPERLITE 1.0. LBL began to get professional market evaluation using the Daylighting Network, supplemented by building industry professionals in North America and abroad.

Daylighting Nomographs. The completed manual and set of nomographs to estimate potential energy and peak-load savings in commercial buildings continued to be a popular professional energy seminar package. More than 400 copies were distributed.

Skylight Design Handbook. Using established research and professional links with the American Architectural Manufacturers Association (AAMA), and their Skylight and Space Enclosure Divison, LBL developed a Skylight Design Handbook/Energy Design Guidelines. In a cooperative project with a professional architectural firm, they converted energy design data into an easy-to-use worksheet the design professional can use to answer basic questions about the skylight design best suited for a building. AAMA will publish and distribute the Handbook to association members and the building industry at large.

Advanced Envelope Design Tool. LBL's attention has turned during the past several years to the development of the next generation of sophisticated hardware and software tools. They envision a tool that relies heavily on imaging technologies, expert-systems software, and design process studies.

LBL has developed an Advanced Envelope Design Tool concept. A schematic illustrating the key factors of this tool concept is shown in Figure 12-6. A computer data base of relevant hardware and software developments was established to assist in identifying and tracking market

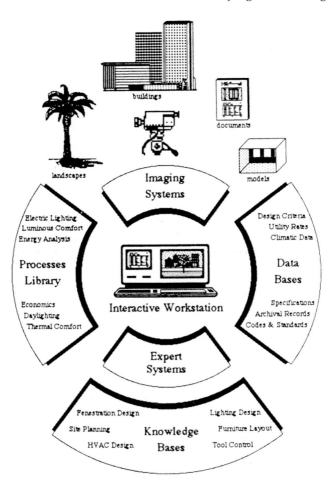

Figure 12-6. Schematic illustrating the key features of the Advanced Envelope Design Tool concept. (XBL 876-2799)

trends. Several major slide presentations were made to organizations in the building professions (AIA, ACED, ASHRAE), and a concept paper published. Planning was initiated to integrate this project with the U.S. Department of Energy's Intelligent Building Design project, to begin in 1988.

Windows and Daylighting
Building 90, Room 3111
Lawrence Berkeley Laboratory
Berkeley, CA 94720

LBL-20087: "Instrumentation for Evaluating Integrated Lighting System Performance in a Large Daylighted Office Building," M. Warren, C. Benton, R. Verderver, O. Morse, and S. Selkowitz, *Proceedings of the National Workshop on Field Data Acquisition for Building and Energy Use Monitoring*, October 16-18, 1985, Dallas, TX.

LBL-21466: "Evaluation of Integrated Lighting System Performance in a Large Daylighted Office Building," M. Warren, C. Benton, R. Verderber, O. Morse, and S. Selkowitz, *Proceedings from the 1986 ACEEE Summer Study on Energy Efficiency in Buildings: Large Building Technologies* (Vol. 3), p. 218, 1986.

LBL-21411: "Field Measurements of Light Shelf Performance in a Major Office Installation," C. Benton, B. Erwine, M. Warren, and S. Selkowilz, *11th National Passive Solar Conference Proceedings*, American Solar Energy Society, Inc. 1986.

LBL-21421: "Spectroscopic and Electrochemical Studies of Electrochromic Hydrated Nickel Oxide Films," P. Yu, G. Nazri, and C. Lampert. *Optical Materials Technology for Energy Efficiency and Solar Energy Conversion V*, Proceedings of SPIE's 1986 International Symposium on Optics and Electro-Optics, Innsbruck, Austria, April 14-18, 1986.

LBL-20347: "The Effect of Daylighting Strategies on Building Cooling Loads and Overall Energy Performance," R. Johnson, D. Arasteh. D. Connell. and S. Selkowitz, *Proceedings of the ASHRAE/DOE-ORNL Conference, Thermal Performance of the Exterior Envelopes of Buildings III*, Clearwater Beach. FL Dec. 2-5, 1985.

LBL-20079: "Window Performance Analysis in a Single-Family Residence." R. Sullivan and S. Selkowitz, *Proceedings of the ASHRAE/DOE Conference, Thermal Performance of the Exterior Envelopes of Buildings III*, Clearwater Beach. FL Dec. 2-5,1985.

LBL-18234: "Transmittance Measurements in the Integrating Sphere," J. Kessel *Applied Optics*. Vol. 25, No. 16, p. 2752-2756.

LBL-20236: "Measured Net Energy Performance of Single Glazing Under Realistic Conditions," J. Klems and H. Keller, *Proceedings of the ASHRAE/DOE-ORNL Conference, Thermal Performance of the Exterior Envelopes of Building Ill*, Clearwater Beach, FL, Dec. 2-5, 1985.

LBL-20348: "Prospects for Highly Insulating Window Systems," D. Arasteh and S. Selkowitz, Presented at Conservation in Buildings: Northwest Perspective, Butte, MT, May 19-22, 1985.

LBL-20080: "Advanced Optical Materials for Daylighting in Office Buildings," R. Johnson, D. Connell, S. Selkowitz, and D. Arasteh, May 1986.

Section VI

Wireless Lighting Control

Chapter 13

Using Wireless Lighting Controls to Reduce Costs, Save Energy, and Enhance Security in Parking Garages

Michael Epling, Robert Best, and Elizabeth Savelle

OVERVIEW

Leveraging advances in wireless technology, existing buildings have the opportunity to capture up to 70% energy savings with advanced lighting controls. This chapter will highlight challenges to retrofitting existing buildings and parking garages for energy efficiency, and examine the energy price drivers and innovations in control technology that enable the economical, non-invasive deployment of lighting controls in existing buildings.

Advanced lighting controls that reduce lighting requirements up to 60% will be used in more than 30% of commercial and industrial buildings in 2020, up from 6% and 12%, respectively, this year.[1] New technologies are rapidly overcoming the limitations of control applications used during building development and reducing the high cost of rewiring facilities for controls. Parking garages in the U.S. also lack lighting controls, and face rewiring challenges and the burden of heightened security and safety concerns. As a result, a majority of garages are designed without lighting controls and are lit 24 x 7.

At Hills Plaza, building management discovered a lighting control technology to limit their exposure to rising energy costs, enable Automated Demand Response (ADR), save energy, and enhance security in the facility's parking garage. The technology used at Hills Plaza delivers a network of wireless occupancy sensors and intelligent light controllers utilizing the ZigBee protocol. The light controllers are installed at the fixture level and respond to wireless occupancy sensor signals in the network, enabling predictive lighting in drive paths and flexible occupancy zones. Designed as a distributed intelligence system, the logic resides in the light controller, eliminating the risk of a single point of failure. An astronomical clock and a current sensor also reside in each light controller, enabling smarter scheduling and energy monitoring at the fixture level. The wireless system is enhanced with real-time energy and data reporting software which integrates with existing automation systems and utility demand response systems.

INTRODUCTION

Hills Plaza, a 3.2 acre multi-tenant, multi-use commercial complex with two popular restaurants in downtown San Francisco south of Market Street, recently completed a comprehensive lighting conversion throughout the parking structure located beneath the commercial buildings and courtyard. The garage covers 186,000 square feet over two levels and offers 314 parking stalls, not counting the area set aside for the owners of 64 luxury condominiums built atop the main commercial building. (Figure 13-1)

Figure 13-1. Hills Plaza

The goals for the parking garage lighting project were to:
• Limit exposure to rising energy costs
• Enable Automated Demand Response
• Save energy and reduce operating costs
• Improve lighting control and light levels
• Enhance security throughout the parking garage
• Implement sustainable operating procedures

DESIGN AND ENGINEERING

To achieve these goals in San Francisco's Union labor market, the Hills Plaza team decided to replace all of the 175 fluorescent light fixtures installed in 1990 when the garage was constructed. Even though the fixtures had been upgraded to T8 lamps and electronic ballasts in 1998, they were simple eight foot four lamp fixtures with no lens and diminishing efficiency. The replacement fixtures were gasketed, vapor tight, and built using an injected thermoplastic design; they also held four F32T8/835/ECO lamps driven by dimming ballasts, but used no metal in the fixture shell.

With no metal in the fixture construction, radio frequency interference that might limit the success of a wireless control strategy was no longer a concern. Consequently, a wireless mesh network based upon the standard ZigBee protocol was selected to communicate between the fixtures and the system vendor's on-site Gateway.

Figure 13-2. Wireless Lighting Control

The Gateway was connected to a secure virtual LAN (vLAN) layered onto the building's Ethernet LAN and linked to the vendor's servers to provide bi-directional communication and control. The possibility of network security issues due to traffic to offsite servers was an acceptable risk mitigated by the secure vLAN configuration.

The anticipated spectral performance of the new fixtures and predictive lighting for the drive paths were both clear improvements over the existing fixtures, allowing the project to move from design to procurement. The new fixtures were ordered and manufactured to include not only the essential electronic dimming ballasts, but also, the occupancy sensor and radio transceivers needed to execute the control strategy. Each fixture was assigned a unique address, and control zones were laid out so movement within the parking structure would automatically trigger strategically placed occupancy sensors to illuminate the pathways to and from guests' cars. Once installed and commissioned, only 45 sensors were required to control the 175 fixtures.

INSTALLATION

With new fixtures in hand, Union staff engineers removed the old fixtures while a vendor-trained Union electrical contractor followed to install the new fixtures one-for-one. The vendor's project management team monitored the progress of the conversion, completing programming and testing of the fixtures while the F32T8 lamps were being conditioned (i.e., operating at full brightness for one hundred hours).

After conditioning, the fixtures were programmed to operate at a maximum of 80% of full brightness, resulting in an immediate savings of 20%. With no activity in a control zone, the fixtures dim to 20% of full brightness while maintaining a minimum illumination of one foot-candle for safety. The installation team also noted the condominium parking area and spaces immediately outside the engineering offices on the lower level were rarely accessed after normal business hours, so a time of use (TOU) schedule was added to turn these fixtures off from 12:30 a.m. to 5:30 a.m. each day. The condominium owners, however, did ask that the minimum lighting level be increased slightly because several owners felt the parking area was too dark. The necessary programming change to raise the standby output level to 40% was completed in minutes!

Results

The vendor recorded energy consumption before and after the lighting conversion; the results are shown in Table 1 and project an annual kWh savings of 41.3%.

The recorded data also demonstrated a higher operating demand on B1, attributed to the activity of two popular restaurants on the lobby level; few patrons choose to park on B2 with space available

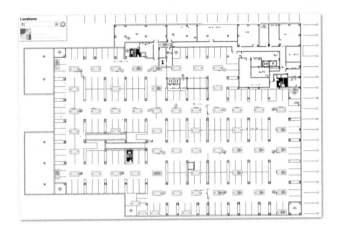

Figure 13-3. Lighting Layout on Parking Level B2 The fixtures are ON when displayed in yellow, and OFF when shown in blue. The occupancy sensors are Green when they are activated and Red when they are not.

kWh/Month	Energy Use		Savings	
	Before	After	kWh	%
Actual July 21 - Aug 20	12,600	7,500	5,100	40.5%
Projected Monthly Average	12,775	7,605	5,170	40.5%
Projected kWh/Year	153,300	90,005	63,295	41.3%

Table 13-1. Energy Consumption and kWh Savings

on B1, and the fixtures on B1 are never turned completely off.

The real time dashboard shown in Figure 13-4 displays the system savings as dollars, kWh, and tons of CO_2 saved since the system was commissioned in early March 2011. The weekly energy history for August 14-20, 2011 is displayed to illustrate the typical consumption pattern for the garage lighting over the course of one week.

Because the lighting fixtures were organized into logical groups, or sublocations, to match their physical locations in the parking garage, the real time contribution to the total load was easily displayed. Figures 13-5a and 13-5b represent snapshots of the three sublocation contributions at two moments during a typical week, mid-afternoon on a weekday and late evening on a Saturday night.

All of these performance metrics can be readily viewed by anyone with Internet access, a browser, and login credentials.

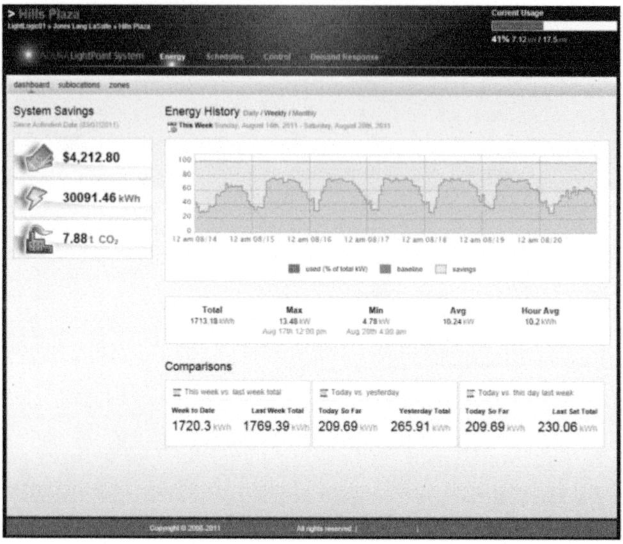

Figure 13-4. Real Time Performance Dashboard

Sublocations

Name	Available	% of Total Available	Current Usage	Usage %	Current Load	Contribution to Total Load
B1 and B1 ANNEX	11.03kW	63%		70.3%	7.76kW	60.9%
B2 CONDOMINIUM SIDE	2.39kW	13.6%		77.6%	1.85kW	14.6%
B2 ENGINEERING SIDE	4.1kW	23.4%		73%	2.99kW	23.5%

Figure 13-5a: Contributions to the total load captured on a typical weekday in mid-afternoon.

Sublocations

Name	Available	% of Total Available	Current Usage	Usage %	Current Load	Contribution to Total Load
B1 and B1 ANNEX	11.03kW	63%		25.9%	2.85kW	57.3%
B2 CONDOMINIUM SIDE	2.39kW	13.6%		41.1%	981W	19.7%
B2 ENGINEERING SIDE	4.1kW	23.4%		29%	1.19kW	23.9%

Figure 13-5b: Contributions to the total load captured on Saturday August 20, 2011 at 11:17 p.m.

ADDITIONAL BENEFITS

The commissioned installation provided several important benefits beyond achieving the short term goals mentioned above. Because the integral electronic components monitor electrical consumption in real time, when a fixture malfunctions or lamps burn out, the web page displaying the status of the fixtures in a zone makes it immediately apparent that a problem has occurred, speeding response and problem resolution.

Additionally, the IP based bi-directional communication between on site Gateway and remote servers enables communication between the local utility and the server/gateway array. During Peak Day Events in Northern California, the communication to the vendor's server and then to the Hills Plaza Gateways and fixtures makes automated demand response (ADR) a real possibility. This lessens the tariff penalties associated with the peak day event. If the end user does not want to automate the demand response or load shedding, the IP based installation makes it simple to monitor and manually adjust the system's status.

FINAL OBSERVATIONS

The success of the lighting conversion and roll out of the wireless control strategy has prompted the Hills Plaza management team to add the process to the building's construction standards. During tenant improvement projects, lighting designers and architects are required to factor the value of the wireless control strategy into the life cycle cost of the proposed build out. Real time energy monitoring, lighting control using occupancy and daylight harvesting sensors, TOU scheduling, and software zoning make the ZigBee enabled wireless control strategy an attractive, cost-effective alternative to conventional hard-wired control upgrades.

References
1.   Adura Technologies at www.aduratech.com XtraLight Fixtures at http://www.xtralight.com/index.php ZigBee Alliance at http://www.zigbee.org/home.aspx

Chapter 14

Energy Savings by Lighting on Demand in Outdoor Lighting with Zigbee Wireless Network

Frank S. Wang

OVERVIEW

Zigbee wireless network adopted by digital dimming electronic ballasts of outdoor hid lamps with an interface (DALI), has been installed on the streets of cities in china. Through a remote control station this advanced lighting on demand system yields significant energy savings. Side-by-side measurements indicated energy savings of 41% to 49% over that of conventional (non dimming) magnetic ballasts. With this information highway in place, a host of other wireless applications can be realized.

INTRODUCTION

Outdoor high intensity discharge (HID) lamps such as high pressure sodium (HPS) and metal halide (MH) lamps have been the untouched lighting area, for more than two decades, as far as the energy saving electronic ballasts (EB) are concerned. Now a digital dimmable EB has been commercially available of up to 400W of HPS and MH lamps in China since year 2000.

The remote management of lighting on demand (LOD) of INDOOR lighting has been in practice with a HARDWIRED system for over 19 years with the availability of dimming EB for fluorescent lamps. But similar remote LOD systems of OUTDOOR HID lighting have not been feasible due to the absence of dimming Electronic Ballasts of HID lamps and NO low cost/low energy WIRELESS communication system. A number of vendors have been exploring the telemetric system (i.e. high frequency control signals through the power supply line) without much success.

By the end of '05 the 2-way wireless communication chips of Zigbee Network had been adopted by our dimming EB of HID lamps. The result is an advanced energy saving system for street (outdoor) lighting by way of remote management of lighting on demand.

The key element in this combination is the development of a special digital accessible lighting interface (DALI) to work between our digital dimming EB of HID lamps and the Zigbee module. Such DALI can convert the control signals from Zigbee central station into proper command actions by EB accordingly. In the mean time, it can convert the performance data of EB into signals to be transmitted to the Zigbee central station when requested, on a REAL TIME basis.

COMMERCIAL INSTALLATIONS

The first test installation of 120 units of HPS lamp fixtures were installed on the street of City of Xuzhou, in Jiangsu Province in the spring of '06. A side-by-side comparison with HPS lamp fixtures of the same wattage, driven by regular non-dimming magnetic ballasts (MB) was conducted with meters. The energy consumed over a couple month period was measured and reported by the Street Lighting Bureau of the city: An energy saving of 49% was observed and documented.

Based on this measured energy saving results, the city authority made the decision to have 1,200 units of such HPS lamp fixtures installed on the streets of its new city district.

The City of Xise of Fujian Province in China also conducted a test installation on its street in the Spring of '07. Again energy savings of 48% were measured with a side by side comparison. This was meant to be the first step of an upgrade project of 30k plus units of HPS lamps of the entire city.

Since then there have been over ten different cities or institutes in China conducting similar side-by-side tests and reported their measured energy savings. Which would range from 41% at City of Haikou, Hainan province to 48% at SungJiang District of Shanghai.

BREAK DOWN OF ENERGY SAVINGS

The energy savings of our LOD in street lighting by Zigbee Network can be broken down into two parts:

(1)   The part contributed by EB vs. MB (made in China).
(2)   The part contributed by dimming between 1:00 am to 5:00 am early morning.

Our electronic ballasts of HPS lamps have been measured to save energy over that of the regular MB in lab as well as by field installations.

A series of field tests, co-sponsored by the Technology Export Program of the California Energy Commission, during a 12-month period from '98 to '00, at three cities in China, Ningbo, Pudong and Tienjin Harbor, were conducted. Energy savings of 20% to 24% were measured. The data and report were presented at energy management conferences in both China and California.

A technical article issued by Philips of China, presented their measured data on the street of the City of Wuxi, between dimming and non-dimming MB. Energy savings of 24% was reported, which was also confirmed with tests at our lab in '08.

Hence the total energy savings of 41% to 49% as measured with our LOD system with Zigbee Network is the summation of Part (1): 20% to 24% and Part (2): 24%.

MAINTENANCE COST SAVINGS

First, there is no need of the routine evening drive-check of any failed lamps on any street sections by the maintenance crew.

Second, preventive maintenance can be dispatched based on the voltage and current readings of any HPS lamp monitored.

Third, savings in spare parts: Based on our field experience during the field tests of '98 to '00, a savings in spare parts cost of 50% to 60% was realized at Ningbo, in terms of replacement of spent MB, damaged igniters and spent HPS lamps.

The most interesting, qualitative observation was that many spent HPS lamps by the MB system were tried on our EB system and found still working well. This was one of the main reasons that convinced the Ningbo authority to conduct that field test.

INCREASE IN LUX OR LUMINS OUTPUT

Another interesting observation at Tienjin Harbor test site was the measurements of LUX of the same HPS lamp driven by MB (made in China) and EB.

The increase in LUX as measured at the same distance from the HPS light source was over 15% to the surprise of the operator. Based on that observation a series of lab tests of integrated lumens by the same HPS lamp driven by MB and EB, had been initiated.

The general conclusion of 10% to 15% more LUX by EB was confirmed at the Lawrence Berkeley National Lab of DOE between '02 to '03. This special physical phenomenon was adopted to further increase energy savings by the dimming operation of LOD with Zigbee Network system at the present stage.

DIMMING UNDER REDUCED LAMP POWER INPUT

The relationship between the lumen's output of an HPS lamp driven by EB and its lamp power input is observed to be nonlinear. Lab measurements indicated that at 50% of the rated lamp power input, the light output is only reduced by 25% of that under 100% of rated power (both driven by the same EB).

The increase of 10% to 15% LUX by electronic ballast, as observed at Tienjin etc, together with this non-linear relationship, enables us to obtain an energy reduction of 50% or more on an HPS lamp driven by EB, on an equivalent LUX basis in China.

Therefore, through dimming by reducing the input power of an HPS lamp, further energy savings can be derived as compared with that driven by regular MB, on an equal LUX basis.

THE Hi-EF HPS LAMP FIXTURE

Based on the measurement of equivalent LUX under reduced lamp power, mentioned here, a series of new HPS lamp fixtures with EB was developed to take advantage of such dimming advantage on a permanent basis. We call such HPS lamp fixtures the Hi-EF HPS lamp fixtures.

For instance, the 250W Hi-EF fixtures are meant to replace (on equal LUX basis) the regular 400W HPS lamp fixtures driven by MB; in turn the 150W Hi-EF fixtures to replace regular 250W HPS lamp fixtures and so on.

This is the first time that an HPS lamp fixture is designed with all the unique features of an EB in mind, such as glass reflector, space for ballast, seal and joint, etc. This shall give the end users a complete package to assure the integrity and consistency of all the excellent energy saving features offered by EB.

ENERGY EFFICIENCY ON THE SUPPLY SIDE

So far the energy savings of a lamp discussed above is confined to the meter side (i.e. the energy consumed as measured by the Watt meter). However in an alternating current (AC) power system there is also the energy (utilization) efficiency of the HPS lamps on the power supply side to be considered. This energy (utilization) efficiency is defined by the power factor (PF) of the lamp system, with a PF of 100% as the highest efficiency.

Because of the inductance of the primary voltage transformer in an MB system, the PF of an HPS lamp can be as low as 35%. Such energy loss due to poor PF is called invisible energy loss (i.e. NOT shown by Watt meter).

To minimize such invisible energy loss, all electric power suppliers in the USA and Western Europe had required a PF of >90% for any electricity consuming system or equipment on end users' premise. Noncompliance shall be subject to a heavy penalty in tariff, called power factor assessment.

Therefore, a compensation capacitor (CC) was required by law, to be an integral part of the MB system to keep the PF at or >90% in the West.

However, such CC is a consuming item with a life cycle much shorter than that of other components of the MB system (except the igniter of MB system). Therefore it is most critical to be able to monitor such power factors *in situ* through a REMOTE WIRELESS system.

MONITORING POWER FACTOR OF OUTDOOR
HID LAMPS WITH A REMOTE WIRELESS SYSTEM

The present LOD system with Zigbee Network can offer, for the first time, a low cost and convenient way to check the power factor of each outdoor HID lamp installed on the lamp post, wirelessly and in real time.

This finger tip type of monitoring of power factor of any outdoor HID lamps should be the best way to reduce such invisible energy loss on the supply side and avoid being penalized by the utility companies in the WEST.

THE IMPACT OF UPGRADING POWER FACTOR
IN THE STREET LIGHTING FACILITIES IN CHINA

Up to the present time, China has no law requiring that the power factor of an HID lamp (< 400W) system be kept at or >90%. In fact, one may safely assume that the PF of the entire inventory of outdoor HID lamps (<400W) in China are less than 40%.

It means that a typical 250W HPS lamp on China streets, driven by a locally made MB system (required no CC or otherwise), would cost the power plant as much as 857W to light that lamp, well over x3 of the rated lamp wattage of 250W.

Such invisible energy loss on the supply side represents the single most wasteful lighting energy practice in China (and in the world). It also points out the most significant area where energy saving lighting device or system should be focused at.

Assuming China shall reach a total of 25 million units of outdoor HID lamps of 400W or less by the year of 2011, the total invisible energy loss by all outdoor lighting facilities due to such poor power factor will be equivalent to the entire power generation capacity of the world largest hydro electric station, the 3-Gorge Hydro-Electric Power Station.

Considering that the 3-Gorge Dam project cost China over 15 billion US$ and over 15 years to build and over 2 million people being moved from their homeland. The social and economic cost is unmatched with any modern infrastructure projects in recent history.

In short, the upgrading of the poor power factor of all outdoor lighting facilities in China shall yield the single most significant impact on the overall power balance (supply vs. demand) of China today.

A SMART SOLUTION THAT
HAS BEEN NEGLECTED FOR SO LONG

One of the main reasons that this power factor problem has been neglected since the very beginning (30-40 years ago) was that it was very difficult (for China or whomever) to check the power factor of each HID outdoor lamp on the lamp post.

Of course, that the utility companies and the city facilities are both government owned identities does not provide any more incentive for the power supplier to address this simple, yet critical problem of invisible power loss.

In a very few cases, where there were compensation capacitors (CC) in the MB system initially, the local maintenance crew would disable, temper or destroy their compensating function once and for all to save trouble and cost in the future. Because there is nobody to check such power factor *in situ*; with or without CC there is no difference on the METER readings (i.e. the electricity cost to the city is the same).

To avoid all such bad habits of local human tempering as well as decades old bureaucracy changes in China, the best cure of this Invisible Energy Loss problem is to adopt a system which can easily monitor all power factors of the HID outdoor lamps from a central station with a laptop.

This is exactly what the present remote LOD Zigbee Network system is offering and IN THE MEAN TIME significant of energy savings on the METER side will benefit the end users.

MANY OTHER USAGES OF SUCH
ZIGBEE NETWORK TO THE CITY GOVERNMENT

The Zigbee Network installed on all street lighting posts in a city shall serve many other functions beside the energy savings on street lighting and the power supply side:

- To monitor the traffic pattern on any busy intersections.

- To manage the city (bus) transportation system according to traffic, loading patterns, route by route, hour by hour (on-peak and off-peak).

- To send signal to hospitals when a street side emergency happens.

- To connect all senior citizens living alone to their medical and health agents when any emergency medical problems happened to any individual (like the current MEDALERT).

- To connect all building HVAC automation systems in a city with a central station (e.g. chain stores, multiple campuses, hospital institutes, office buildings etc).

- To manage similar functions of several cities via GPRS or internet or satellite communication system.

WHY ZIGBEE NETWORK SYSTEM WAS NOT WIDELY APPLIED IN THE USA?

The shorter distance (100 meters) signal transmission of Zigbee Network naturally limits its applications to only indoor environment, such as residential, apartment, office building (refer to the Zigbee Conference in Chicago, May 2007).

To expand the usages of Zigbee Network over a wide area, one must cover a city or a region of a few cities with many Zigbee transmission points. This involves the cost of physically laying out such a network in a wide area. This is a tall order of business investment. Question: who is going to pay for it?

In the USA, utility companies are heavily relied upon to invest into this network in a city or his service territory with residential automatic meter reading, energy management system etc.

This is not going to work well when there is an inherent conflict of interest between power consumers and power suppliers. Without getting around this bottle neck Zigbee Network will not be as valuable and versatile as it should be.

CONCLUSION

We believe that the remote management of LOD in street (outdoor) lighting with Zigbee Network, pioneered by us in China is a win-win situation to all parties involved:

- End users, less cost on meter side to the city;

- Power suppliers, less invisible power loss, same power plant can serve more customers, NO need to build new power plants etc;

- Society as a whole, less pollution from coal fired power plants, reduced carbon emission and reduced consumption of natural resources.

- A free citywide wireless network to bring out all other extended applications mentioned above and more.

Therefore we may conclude that the benefits of the Zigbee Network system applied to street lighting for the sake of energy saving CAN go far beyond the cost reduction on the meter side and the savings in invisible power loss. It is the LOWEST COST approach toward a digital city management of the future.

References
1. "Energy Savings with High Pressure Sodium lamps on the street of City of Xuzhou" Report by the Street Lighting Bureau of City of Xuzhou, Jiangsu, China.
2. "Energy Savings with High Pressure Sodium lamps on the street of City of Xise, Fujian, China
3. "Field Measurement on Energy Savings of High Pressure Sodium lamps driven by Electronic Ballasts and Magnetic Ballasts at Ningbo, Pudong, Tianjin, China," Paper presented at Western Energy Management Congress, June 2002
4. "Zigbee Alliance Addressing Energy Efficiency," presentation at LightFair International, Las Vegas, May 29, 2008 by Bob Heile
5. "Zigbee in Energy: Innovation, Collaboration, Conservation" presentation at Connectivity Week, Chicago, May 22-24 2007 by Ed May, Itron
6. "Zigbee in Commercial Buildings," presentation at Connectivity Week, Chicago; by Bob Gohn, Ember Corp. May 24 2007

Section VII

Case Studies

Chapter 15

Mesopic Outdoor Lighting:
Can a Light Source Tuned to How Humans See at Night Save Energy?

Peter M. Morante

PROJECT OBJECTIVE

Can a white light source tuned to how humans see at night under low light levels—one with lower wattage and photopic light output—replace a high-pressure sodium (HPS) street lighting system and still provide equal or greater perceptions of visibility, safety, and security? If so, when and where should this lighting system be used?

The simple answer is YES. How the mesopic vision system works and the results of experimentation are discussed below.

THE SCIENCE BEHIND MESOPIC VISION AND LIGHTING

The human vision system has two types of receptors in the retina, cones and rods, to transmit visual signals to the brain. The current system of photometry to determine the amount of light needed to perform a task, regardless of the time of day or lighting conditions, is based on how the eye's cones function. Cones are the dominant visual receptor under photopic (daylight) lighting conditions. Rods function primarily under dark (scotopic) conditions. Under mesopic lighting conditions, which are typically found outdoors at night under low lighting conditions produced by street and parking lot lighting, a combination of cones and rods perform the vision function. Therefore, outdoor electric light sources that are tuned to how humans see under mesopic lighting conditions can be used to reduce the luminance of the road surface while providing the same or better visibility. This light source must account for how both the cones and rods in the eye see. Light sources with shorter wavelengths, which produce a "cooler" (more blue and green) light, are needed to produce better mesopic vision.

Figure 15-1 shows the visible spectrum with photopic vision peaking at 555 nanometers (violet graph) and scotopic vision peaking at 507 nanometers (green graph). Mesopic vision is a combination of both photopic and scotopic and will peak someplace between the peaks of photopic and scotopic depending on the illuminance value. The higher the illuminance value, the closer the mesopic response will be to the photopic graph. A white light source will have less energy savings potential as light levels increase. With lower illuminance, the mesopic peak will move toward the scotopic peak. Greater energy savings potential exists when using white light sources at lower light levels.

PREDICTING THE AMOUNT OF
LIGHT NEEDED UNDER MESOPIC CONDITIONS

Current photometry underestimates the effectiveness of lamps with relatively more short-wavelength output at mesopic light levels (< 10 foot-candles). The unified photometry system developed by the Lighting Research Center (LRC) can more appropriately evaluate the effectiveness of lamps with various spectral power distributions (SPD) by providing "unified" luminance according to the light levels to which human eyes adapt.

The LRC developed the unified photometry system based on a series of laboratory studies. Simulated driving studies verified the validity of the fundamental findings but underscored the fact that light level as well as target contrast and size affect off-axis detection. Therefore, the visual performance differences between metal halide (MH) and high pressure sodium (HPS) lamps can be even larger than would be predicted by the unified photometry system alone. A recent field study to examine target detection by subjects driving along a closed track found that targets illuminated by MH lamps can be more quickly detected than those made visible by HPS lamps (Akashi and Rea 2002). The results dramatically underscored the benefits of the unified photometry system.

Table 15-1 shows photopic illuminance and relative electric power required to obtain criterion levels of off-axis visual performance when illuminated by various SPD light sources. As the light level decreases, the performance of HPS lamps, relative to other sources,

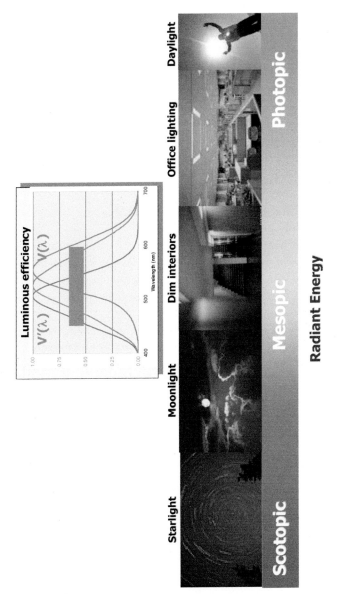

Figure 15-1. Mesopic profile

is reduced. Conversely, MH and fluorescent lamps, which have more short-wavelength components, reduce their relative power requirements to meet criterion visual performance levels.

The unified photometry system is used in part to develop criteria for the HPS replacement lamp. The system provides the equivalent mesopic luminance for lamps of differing SPDs that will produce equivalent visual performance under low light (mesopic) conditions. The ratio of a replacement lamp's scotopic luminance to photopic luminance (S/P ratio) is used as one of the variables to determine the necessary unified luminance flux equivalent to the HPS lamp. The other variable is the required luminance of the road surface.

TAKING THE MESOPIC RESEARCH
INTO THE REAL STREETLIGHT WORLD

The LRC has conducted four demonstrations and evaluations where HPS street lighting was replaced with various other lighting technologies all of which had shorter wavelength (more blues and greens) light. Table 15-2 depicts the original HPS and its replacement

Table 15-1. Photopic illuminance and relative power required to obtain the same brightness perception and visibility of spaces and objects illuminated by various SPD lamps

Light source	S/P ratio*	0.6 cd/m²		0.3 cd/m²		0.1 cd/m²	
		E (lx)**	Relative power***	E (lx)	Relative power	E (lx)	Relative power
400 W HPS	0.66	26.9	100%	13.5	100%	4.5	100%
1000 W incandescent	4.41	26.9	833%	10.5	648%	2.6	478%
3500 K fluorescent	1.44	26.9	130%	10.4	100%	2.5	73%
400 W MH	1.57	26.9	119%	10.0	88%	2.4	63%
5000 K fluorescent	1.97	26.9	130%	9.0	87%	1.9	57%
6500 K fluorescent	2.19	26.9	130%	8.5	82%	1.8	52%

* S/P ratio: The ratio of scotopic lumens to photopic lumens of each lamp
** E: Illuminance measured in lux (lx)
*** Relative power (%) normalized to HPS

light sources. The evaluation results for all four demonstrations are similar. The results of the Groton, Ct. demonstration where 100 watt HPS were replaced with 55 watt induction (electrodeless) light sources are presented here.

EVALUATION RESULTS

Lamp and light fixture information for Meridian Street for both the initial HPS street lighting and for the induction lighting demonstration are listed in Table 15-3.

Table 15-2. Streetlight demonstration light sources (before and after)

Location	Initial Lighting		Replacement Lighting	
	CCT	Type & wattage	CCT	Type & wattage
Easthampton, MA	2100K	HPS, 70W	6500K	Fluorescent 50 W
Austin, TX	2100K	HPS, 250 & 100 W	4100K	Fluorescent 100 W
Groton, CT	2100K	HPS, 100 W	4000K	Ceramic MH, 70 W
Groton, CT	2100K	HPS, 100 W	6500K	Induction, 55 W

CCT: correlated color temperature

Table 15-3. Streetlight fixture information

	Initial HPS Street Lighting	Induction Demonstration Street Lighting
Lamp Type	HPS	Induction
Lamp Wattage	100 watts (118 W with ballast)	55 W (including power for driver)
CCT	2100 K	6500 K
Mean Lumens	8460	3300*
Lamp Life	30,000 hours non-cycling	60,000 hours
Fixture Type	Cobra Head	Cobra Head
Number of Fixtures	12	12
Light Distribution Type	Type II	Type II
Cutoff Classification	Cutoff	Cutoff
Mounting Height	25 feet	25 feet
Avg. Illuminance	8.72 lux*	2.69 lux*

* From LRC test and measurement results

Figure 15-2 illustrates the results of comparing the induction and HPS street lighting at the Groton demonstration site. Graph bars tracking to the right toward the positive end of the scale indicate agreement with the survey statement, while bars tracking to the left toward the negative end of the scale indicate disagreement with the statement.

The results depicted in Figure 15-2 show a strong preference for the induction lamp at 6500 K CCT. Survey respondents indicated that they felt safer and could see better with the 55-watt induction lamp at 6500 K CCT than with the 100-watt HPS at 2100 K CCT.

Possibly the best comparison of visibility, brightness, and color rendering can be seen in Figures 15-3 and 15-4. These photographs were taken on the same section of roadway with HPS lighting and after the installation of the induction lighting. The photos exposures are similar with regards to a combination of F-stop, exposure time, and ISO speed. The resident with the white car had it parked in identical positions for both photos, even though Figure 15-3 was taken on April 27, 2007 and Figure 15-4 on August 27, 2007. Please examine closely the details of the white car in both photos and the visibility and color of the vegetation. It is clear from these photos why residents' perceptions of visibility, brightness, and color rendering were higher with the induction lighting at 6500 K CCT. However, one must take into account that Figure 15-3 was taken with some fog moving into the area.

ENERGY SAVINGS AND ECONOMIC ANALYSIS

Replacing a 100-watt HPS light source (118 watts with ballast) with a 55-watt induction light source saves considerable amounts of energy. Assuming 4,160 hours of operation per year, the 100-watt HPS system will use 491 kilowatt-hours (kWh), compared to the 55-watt induction system at 229 kWh. This is a 53% reduction in annual energy use for the induction system.

Lifecycle cost analyses examines the capital cost of the street lighting made in Year 1 and all annual energy and "other" costs and maintenance costs as they occur over the 27-year life of the streetlight fixture. "Other" costs include company overheads, taxes, etc. For this analysis, there is no salvage value or disposal costs at the end of life. A 6% annual discount rate was used to determine present value of all

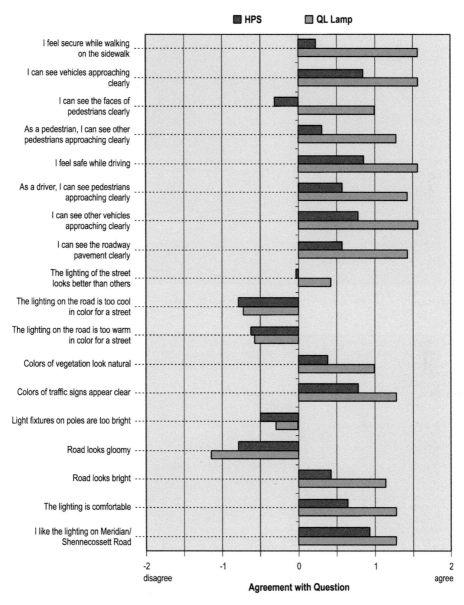

Figure 15-2. Streetlight comparison on Meridian Street: HPS and induction (QL)

Figure 15-3: Initial 100 W HPS street lighting

Figure 15-4. Induction street lighting

costs. Table 15-4 shows the lifecycle cost of each streetlight system for the Groton demonstration site.

Table 15-4. Streetlight systems lifecycle costs

	HPS	Induction
Capital Cost	$268.00	$550.00
Lifecycle Energy Cost	$1005.06	$468.45
Lifecycle Maintenance Cost	$218.63	$190.34
Lifecycle "Other" Cost	$340.96	$340.96
Total Lifecycle Cost	$1832.65	$1549.75

- Residents' perceptions of visibility, safety, security, brightness, and color rendering improved considerably under the induction and ceramic street lighting compared to their perceptions with the HPS street lighting.

- The energy-saving potential compared to HPS is substantial (up to 50%) with the use of a lighting source tuned to how humans see under low light levels.

- The unified photometry system developed by the LRC is a good predictor of visual performance under different lamp scotopic-to-photopic ratios and varying photopic luminance values.

The full report can be viewed at www.lrc.rpi.edu/researchareas/pdf/GrotonFinalReport.pdf

Chapter 16

Assessment of the Lighting Systems For the Oil Sector Complex Building in Kuwait— Before and After Commissioning

Dina AlNakib, CLEP, and Dr. Ali Hajiah

OVERVIEW

Achieving visual comfort level is very important in artificially lit areas. Adequate design of the lighting systems is a key factor for increased personnel productivity. The cooperation between the lighting systems designer and the interior designer is important for optimum lighting design and visual comfort. This chapter demonstrates the effect of furnishings on the lighting systems in the Oil Sector Complex (OSC) building in Kuwait which was commissioned in 2007. Continuous monitoring of the lighting systems in the complex was performed before and after commissioning. Walkthrough surveys to evaluate the illumination levels and the effect of daylight in the building were conducted. The occupants' visual comfort was investigated and further recommendations to improve the lighting systems operation scheduling and energy consumption in the building were provided. The effect of using energy conservation measures such as occupancy sensors was analyzed. Recommendations to improve the lighting systems and solutions to improve the illumination levels of the offices were also proposed.

Key Words: Visual comfort, illumination levels, occupancy sensors

INTRODUCTION

The Oil Sector Complex (OSC) Building, located in Shuwaikh, Kuwait City, is one of the most modern buildings in Kuwait. Housing both the Kuwait Petroleum Corporation (KPC) and the Ministry of

Oil (MOO), the complex was commissioned in 2007 containing the most recent lighting systems and the double glazed windows surrounding the building provide a sufficient amount of daylight inside the building.

The KPC management decided to assess the indoor air quality along with the lighting systems in the OSC prior to the relocation of their employees into this new building. Another assessment of the lighting systems was conducted after commissioning as well.

This chapter illustrates the methodology and results of the assessment of the lighting systems in the OSC building before and after commissioning. A comparison between the two cases and recommendations to improve the lighting systems of the OSC in terms of quality and quantity are also provided.

BACKGROUND

Achieving a visual comfort level is an important factor in artificially lit areas. This will increase the occupants' productivity and reduce eye strain as well. Providing the right amount of light while performing a task is vital to achieve that task successfully. Kuwait's Ministry of Electricity and Water (MEW) developed some guidelines within the Energy Conservation Code of Practice for the optimum illumination levels for different applications depending on the type of the visual task[1]. Utilization of daylight, whenever possible, will further enhance productivity and reduce the consumed energy. In addition, the ability of the light to reflect from the surfaces must always be taken into consideration since dark surfaces absorb light and light surfaces reflect it. Therefore having light colored surfaces will enhance the lighting systems performance in providing adequate lighting to the different areas. The application of energy-efficient luminaires is also crucial since they have the ability to distribute lighting efficiently in addition to deflecting the heat generated from the lamp.

A comprehensive assessment of the lighting systems in the OSC building was conducted based on the continuous monitoring of the lighting systems performance before and after commissioning.

The main feature of the new OSC building is the large glazed area which allows daylight to penetrate the building throughout the day. As for the interior of the building, the finishes of the surface in the main corridors and hallways are of light color. Conversely, the carpets and partitions in the offices are dark, which absorb the light and make the area seem dark. It should be noted that the types of lamps and luminaries used in the OSC building are efficient and of the latest technologies.

LIGHTING SYSTEMS DESCRIPTION

The lighting systems used in the OSC building are efficient. The lamps are provided by an internationally accredited supplier, energy efficient and of the latest technologies. The lamps' color is daylight, which is recommended for office buildings.

The corridors are lit with energy-efficient compact fluorescent lamps (CFLs) housed in recessed luminaires with aluminum reflectors. The offices are lit with a mixture of CFLs downlights and direct-indirect pendant fluorescent tubes luminaires, shown in Figure 16-1. Direct-indirect fixtures are recommended in offices with extensive use of computers because they reflect half of the light towards the ceiling therefore avoiding glare on computer screens[2]. All fluorescent lamps use electronic control gears (ECGs).

The control of the lighting systems is achieved through the Building Automation System (BAS). Each area is remotely controlled by the BAS operators through its allocated distribution board (DB). There are no manual controls in the offices of the OSC building except for the meeting rooms which have dimmer controls. This leads to unnecessary waste of lighting energy since the lights in

Figure 16-1. The lighting system in an office area.

some unoccupied offices are switched on while there are no occupants inside them. In addition, since the building is well-glazed, some areas with adequate daylight have their lights switched on when not needed.

METHODOLOGY

A walkthrough survey was conducted before commissioning. During the walkthrough survey the lighting systems were investigated, the types of lamps were verified and the illumination levels were measured using a light meter. The measured illumination levels were then compared with the illumination levels recommended by the Ministry of Electricity and Water (MEW) Energy Conservation Code of Practice.

Another walk-through survey was conducted after commissioning of the complex. The illumination levels for some selected locations were measured after the building was occupied. Interviews with the building's operators were conducted to study the operation and scheduling of the lighting systems.

As for the assessment of the employees' visual comfort, a questionnaire was distributed to the employees of the complex to evaluate any complaints and provide solutions for enhancing the occupants' visual comfort to perform their work efficiently.

EVALUATION OF THE LIGHTING SYSTEMS
BEFORE COMMISSIONING

The illumination levels were measured in different areas in the building before commissioning. In general, the illumination levels in well-glazed areas such as hallways, corridors and lift lobbies, reached as high as 4000 lux. The recommended illumination levels specified for these areas by MEW is between 100-150 lux[1]. The illumination levels in the offices varied between 270-500 lux with the task lights switched off; and between 470-900 lux with the task lights switched on. The recommended illumination levels for offices are between 300-500 lux[1].

Details of the average measured illumination levels in different areas of the OSC building and the recommended levels are listed in Table 16-1.

As indicated in the Table 16-1, the illumination levels in the OSC building are higher than those specified by MEW; this is due to the excessive natural lighting penetrating the building.

In order to assess the excess amount of artificial lighting used in glazed areas, a procedure that measures the artificial lighting contribu-

Table 16-1. Average Measured and Recommended Illumination Levels in Different Furnished Areas of the OSC Building.

Location	Illumination Levels (Lux)	
	Measured	Recommended
Auditorium	280	100
Corridors (with daylight)	1280	-
Corridors (without daylight)	310	100
Kitchenette	560	500
Lift Lobby	840	150
Main Entrance	2130	150
Meeting Rooms	450	300 – 500
Offices	380	300-500
Offices (Task light on)	550	300 – 500
Reception Desk	820	500
Rest Rooms	340	100
Stairways (Emergency)	890	150

tion in the glazed areas of the building was carried out. The procedure included measuring the illumination levels in glazed areas twice, once with the lights switched on, and another with the lights switched off. The following equation was used to measure the percentage of the contribution of the artificial lighting in the illumination level:

$$PCAL = [(IL_{on} - IL_{off})/(IL_{on})] \times 100 \qquad (1)$$

Where:

PCAL = Percentage of the Contribution of the Artificial Lighting.
IL_{on} = Illumination level with lights switched on.
IL_{off} = Illumination level with lights switched off.

Some of the well-glazed areas in the OSC building that have artificial lighting switched on are: the main entrance, lift lobbies, basement corridors with skylights, and some offices, as shown in Figures 16-2, 16-3 and 16-4. The measurements of the illumination levels and the percentages of the contribution of the artificial lighting for some selected locations in the OSC building are listed in Table 16-2.

According to Table 16-2, the contribution of the artificial lighting in well-glazed areas is between 26% and 60%. For example, the contribution of the artificial lighting in the lift lobby is 26%, while 74% of the light is lit by natural lighting. This means that 74% of the artificial lighting is wasted[2].

Figure 16-2. The lift lobby.

Figure 16-3. The main entrance and the reception desk.

EVALUATION OF THE
LIGHTING SYSTEMS AFTER COMMISSIONING

Upon the KPC's request, another assessment of the lighting systems was performed after commissioning. The request was due to the fact that some employees in the OSC building were complaining of inadequate lighting. Therefore, another detailed walkthrough survey was conducted to evaluate the lighting systems. In order to find out the types of complaints and their locations, a questionnaire was submitted to the employees and mainly inquired about the adequacy of the lighting systems and the illumination levels in the complex.

Figure 16-4. Office with glazing.

Table 16-2. Percentages of the Artificial Lighting Contribution in the Illumination Levels for Different Glazed Areas.

Location	Illumination levels (Lux) Lights Switched On	Lights Switched Off	Artificial Lighting Contribution (%)
Basement	1210	480	60
Main Entrance	960	630	34
Lift Lobby	1600	1190	26
Offices	790	400	49
Reception Desk	820	570	30

Walkthrough Survey

Although the lighting systems are efficient and the illumination levels were high compared to MEW recommendations, the interior design of the building was planned without taking into consideration the layout of the luminaires. Conversely, the lighting designer did not put into consideration the dark colored partitions and carpets which have a great effect on the perception of the illumination (Figure 16-5).

Measurements of the illumination levels in some areas were taken and compared with the measurements conducted before the commissioning of the building. The results are shown in Table 16-3.

The illumination levels in the offices have increased because some employees complaining of inadequate illumination levels have increased the fluorescent lamps' rating from 49W to 80W. The illumination levels in partitioned desks with the task lights on has slightly decreased because the measurements were taken before the desks were

Figure 16-5. Distribution of the lighting systems and the partitions.

Table 16-3. Comparison between Illumination Levels in the Facility before and after Commissioning.

Location	Illumination Levels (Lux)		
	Before Commissioning	**After Commissioning**	**Recommended**
Auditorium	280		100
Corridors *	1280	1180	-
Corridors **	310	290	100
Kitchenette	560	560	500
Lift Lobby *	840	800	150
Main Entrance *	2130	700	150
Meeting Rooms	450	450	300
Offices	380	540	300-500
Offices (Task light on)	550	530	300-500
Reception Desk	840	520	500
Rest Rooms	340	340	150
Stairways *	890	890	150

   * With daylight
   ** Without daylight

occupied, any object added on the desks can absorb the light.

A considerable decrease in the illumination levels was noticed at the OSC building main entrance and the reception desk because of the furnishings and the addition of a large picture at the glazed area (Figure 16-6 a and b). Simultaneously, the corridors without daylight have been affected with the decrease in the illumination levels due to the dark colored carpets. In addition, due to the labor work dur-

Figure 16-6. The main entrance. (a). Before commissioning. (b). After commissioning.

ing the furnishing of the building, dust build-up was noticed on the luminaires, which has affected the light output.

A preliminary solution for the occupants who perceived their offices as dark was to add task lighting. Although increasing the lamps' ratings was more popular, it was not an energy efficient solution since the lighting energy consumption will be almost double than it used to be.

Similar to the KPC building, the design and distribution of partitions in MOO building is not in line with the light distribution as displayed in Figure 16-7. Glare on the computers' monitors is mostly found in secretaries' offices because of the offices' orientation where the monitors face the windows. Some partitions have glare on the monitors because of the luminaires positioned above them[4].

Figure 16-7. Distribution of partitions within the MOO offices.

A Study of the Operation of the Lighting System

The operation schedule for the lighting systems in the OSC building was studied with the help of the OSC's BAS operators. The lights are fully controlled by the BAS through the distribution boards. The lighting schedule does not take into consideration the weekend days. Table 16-4 shows the lighting schedule as provided by the BAS operators[4].

Table 16-4. Lighting Schedule for the OSC Building

Location	ON	OFF
Aviation Light	5:30 PM	6:30 AM
MOE & KPC (Toilet, inner & outer office)	6:00 AM	12:00 AM
External Light	5:45 PM	6:15 AM
Bollard Light	5:45 PM	10:00 PM
Car Park Light	5:45 PM	6:15 AM
Level 19 KPC (VIP)	24 hours	
All Building Corridor Light	24 hours	
KPC, MOE & Center Core	24 hours	

A Study of the Survey Submitted to the OSC Employees

A questionnaire was distributed to the OSC employees to find out the areas in the building where the lighting systems were not adequate according to the occupants. An assessment of the questionnaire was performed and the survey revealed that 35% of KPC's employees complained about the lighting in the building[3]. However, when the team inspected their offices, many of them have increased the lamps ratings from 49W to 80W, which provided more illumination. As for the MOO employees, 38% found the lighting not adequate. Dark perception in offices was due to the dark surface colors.

The employees' complaints in both MOO and KPC buildings were due to the following reasons:

- Dark perception due to dark colored surfaces as shown in Figure 16-8.
- The design and distribution of partitions is not in line with the light distribution.
- Glare on the computers monitor is mostly found in secretaries offices because of the offices orientation where the monitors face the windows.
- Some partitions have glare on the monitors because of the luminaires positioned above them; however, the screen could not re-located due to difficulty in positioning the electrical cords as shown in Figure 16-9.

Continuous Monitoring of the Illumination Levels
Using Data Loggers

Data loggers that continuously measure the illumination levels were installed in some selected areas in the OSC building from March 2007 until March 2008. The measurements were continuously downloaded and the results were analyzed. Since the lighting systems are not affected by weather seasonal changes, there were no significant variations in the collected data throughout the year[4,5]. Figures 16-10 through 16-13 show profiles of the collected data for the illumination levels measured by the data loggers in some selected areas for a week day and a weekend day for both KPC and MOO buildings.

Figure 16-8. Dark surfaces absorb the light.

Figure 16-9. A computer with glare on the monitor.

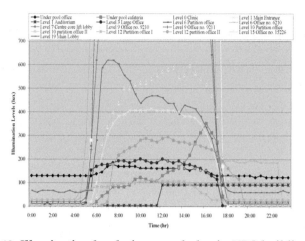

Figure 16-10. Illumination levels for a week day in KPC building (Tuesday, September 18, 2007).

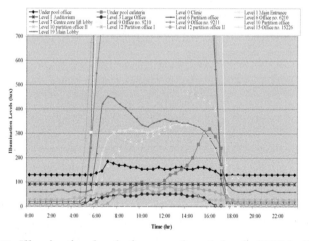

Figure 16-11. Illumination levels for a week-end day in KPC building (Friday, September 21, 2007).

The figures above demonstrate that the illumination levels in the KPC offices that have no glazing vary between 200-500 lux compared to the MEW recommended levels of 300-500 lux. High illumination levels in the main entrance lift lobby and level 19 main lobby in addition to some offices are due to the excessive amount of daylight.

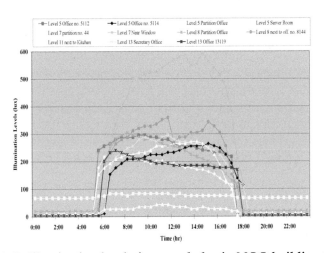

Figure 16-12. Illumination levels for a week day in MOO building (Tuesday September 18, 2007).

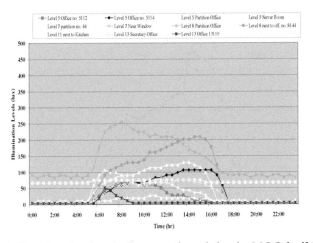

Figure 16-13. Illumination levels for a week-end day in MOO building (Friday September 21, 2007).

Further measurements were carried out using a light meter to verify and investigate the low values of illumination levels downloaded from the data loggers in some areas of the OSC building. Table 16-5 shows a comparison between the illumination levels measured by a light meter.

Table 16-5 indicates that the actual illumination levels measured using a light meter are within the recommended illumination levels provided by the MEWater which are between 300-500 lux. The difference between the readings of the downloaded data using data loggers and the measured data using a light meter is due to the fact that the actual measurements from the light meter are taken horizontally at the task level, on the desks for offices and on the floor for corridors and lobbies. While the data loggers, which are placed on the walls, measures vertically and shading affects the measured readings[6].

Table 16-5.
Illumination Levels Recorded by
Data Loggers and Measured by a Light Meter.

Location	Illumination levels (lux)	
	Data Logger	Light Meter
Clinic	75	350
Level 3: Large office	200	420
Level 12: Partition office	100	440
Level 15: office 15226	170	400

FEASIBILITY STUDY FOR OCCUPANCY SENSORS

To study the feasibility of using occupancy sensors in the OSC building, an occupancy sensor was installed in an office at the KPC. A data logger to measure the illumination levels in the office was installed as well. The measured illumination levels were used as an indicator that the lights in the office are switched on or off. The savings and economic feasibility of the application were evaluated.

The data for the illumination levels in the selected office were collected for a week before and a week after the installation of the occupancy sensor.

The data indicate that the lights in the office were switched off at 7:30 PM through the building automation system (BAS), while with the occupancy sensor, the lights were switched off after the departure of the employee around 3:00 PM. There is a great potential of savings with the occupancy sensors due to the short operation hours of the lighting systems.

To perform the economic feasibility of the application of occupancy sensors in the OSC building, the following parameters were taken into consideration:

- The working hours are between 7:30 AM and 3:00 PM, Sunday to Thursday.

- The lighting system is controlled through the BAS with operation hours between 6:00 AM and 6:00 PM.

- The total days off within a year are around seven weeks which include official holidays, sick leaves and annual leaves.

- Cost of an occupancy sensor including installation charges is $ 150.

- Cost of electricity for the consumer is 0.6 cents/kWh.

Table 16-6 shows the load, energy consumption and energy cost before and after the use of occupancy sensors.

Table 16-6.
The Analysis of Using the Occupancy Sensor in One Office in KPC Building.

	Before Implementation	After Implementation
Days on/yr	261	225
Days off/yr	104	140
Operation hours/yr	3132	1688
Office Lighting Load (W)	98	98
Consumption (kWh/yr)	307	165
Cost of Energy ($/yr)	1.8	1
Energy savings (kWh/yr)		142
Percentage of savings		46%

Table 16-6 shows that 46% savings in energy consumption can be obtained due to the use of an occupancy sensor in one office in the KPC building.

To perform the economic feasibility of the application of occupancy sensors, the following parameter is defined[7]:

$$\text{Cost of conserved energy (CCD)} \left[\frac{\text{KD}}{\text{kWh}} \right] = \frac{\text{annualized investment}}{\text{Annual electricity saved}} \quad (2)$$

where,

$$\text{Annualized investment} = \text{total investment} \times \frac{d}{1 - (1 + d)^{-n}} \quad (3)$$

and

    d=   the discount rate
    n=   the lifetime of the conservation measure (yr).

This parameter allows comparison between different measures and reflects the cost of saving 1 kWh. Moreover, the CCE incorporates the lifetime of the conservation measure as well as the time value of money through the use of the discount rate. The discount rate used in the analysis is that determined by the Central Bank of Kuwait (CBK), which was 5.75%[8] at the time of the analysis.

Assuming the lifetime expectancy of the occupancy sensor to be 20 years, and the total investment of US $ 150 for the sensor, then:

Annualized investment =150 x [0.0575/(1 − (1.0575) − 20)]
                            = 12.8 $/yr
CCE = 12.8/142 = 0.09 $/kWh = 3 cents/kWh

The annual cost for installing occupancy sensors for the next 20 years is 3 cents/kWh, whereas currently the price of electricity is 0.6 cents/kWh; this means that an additional 2.4 cents/kWh will be paid yearly for installing occupancy sensors. Although the Kuwaiti government, through MEW which is highly subsidizing the price of electricity, is the main beneficiary for this measure, it is not economically feasible for the OSC facility to install occupancy sensors in the offices at the present time. However, the corridors, in which the lights are continuously on, and the meeting rooms are good locations for the installation of occupancy sensors.

To study the feasibility of installing occupancy sensors in the meeting rooms, further assumptions were considered. Assuming that the meeting rooms of the OSC building are used for three hours daily, Table 16-7 shows the impact of using occupancy sensors in one meeting room in the OSC building.

Table 16-7 indicates that 75% savings in the energy consumption in meeting rooms can be obtained by installing occupancy sensors. Using the same parameters above, the feasibility of installing occupancy sensors was evaluated using annualized investment of 12.8 $/yr as follows:

$$CCE = 4.3/688 = 0.018 \text{ \$/kWh} = 1.8 \text{ cents/kWh}$$

In spite of the vast savings obtained in energy consumption, an additional 1.2 cents/kWh will be paid annually for installing the occupancy sensor. This can be justified if we consider the increase in lamp life and the reduction in the building's cooling load[6].

Table 16-7.
The Effect of Using an Occupancy Sensor in a
Meeting Room in OSC Building.

	Before Implementation	After Implementation
Operation hours/yr	3120	780
Meeting Room Lighting Load (W)	294	294
Consumption (kWh/yr)	917	229
Cost of Energy (KD/yr)	5.5	1.4
Energy savings (kWh/yr)		688
Percentage of savings		75%

CONCLUSIONS AND RECOMMENDATIONS

Although the lighting systems in the OSC buildings are very efficient and of the latest technologies and the types of fixtures used are as recommended for office buildings, the incorporation of furnishings and partitions was not in line with the design of the luminaires. Therefore, the cooperation between the lighting designer and the interior designer prior to the interior design of any building is vital so that both lighting and interior design will complement each other to provide adequate illumination levels and avoid any visual discomfort for the building's occupants.

The following suggestions are recommended to enhance and improve the lighting systems performance in the OSC building:

• Reorganizing the desks in the offices with low illumination levels and aligning them with the luminaires above.
• Enhancing the illumination levels in the offices by adding task

lights with energy efficient lamps and avoid increasing the fluorescent lamps rating in the luminaires.

- Relocating the computers' monitors in the partitions areas to avoid glare. This requires drilling the desks to rewire the computers circuits.

- When refurnishing the interior of the OSC building, it is recommended to use light colored surfaces and carpets to enhance the illumination levels. It is also important to relocate partitions and align them with the luminaires above to avoid glare and improve the illumination level in the work space.

- Rescheduling the lighting system operation according to the occupancy pattern of the building. Switching off the lights in the offices and non-critical areas after 4:00 PM is recommended. A memo can be circulated to all employees indicating that in case an employee wants to stay in office after working hours, they must inform the BAS operators to switch the lights on in the required area.

- Using occupancy sensors in offices is not economically feasible at the present time due to the subsidized price of electricity in the country. However, it is recommended to use occupancy sensors in low circulation areas such as meeting rooms and restrooms.

- Rewiring the lighting systems and adding manual controls in well-glazed offices with high daylight contribution are recommended to avoid wasted lighting energy.

- Maintenance of the luminaires is recommended. Periodical cleaning of the lighting fixtures can remove dust build-up, enhance the illumination level, increase lamp life and improve visual comfort in the OSC building.

References

1. MEW. 1983. Energy Conservation Program, Code of Practice, MEW/R-6, First Edition. Ministry of Electric and Water, Kuwait.
2. Khan, A.R., A.E. Hajiah, A. Ramadan, L. Al-Awadhi, B. Gevao, G. Maheshwari, D. Al-Nakib, F. Alghimlas. 2006. Indoor Air Quality and Environmental Assessment Study for the New Corporate Oil Sector Complex. Final Report. KISR 8078.
3. Alghimlas F., A.E. Hajiah, D. Al-Nakib, R. Alasseri. 2007. Provision of Continuous Monitoring & Reporting Services—Indoor & Outdoor Air Quality of Oil Sector Complex Building (Part II). Assessment of HVAC and Lighting System Performance and Indoor Air Quality of the Oil Sector Complex Building. Progress Report No. 1. KISR 8805.
4. Al-Nakib, D., A.E. Hajiah, F. Alghimlas, M. Sebzali, Alasseri. 2007. Provision of Continuous Monitoring & Reporting Services—Indoor & Outdoor Air Quality of

Oil Sector Complex Building (Part II). Assessment of HVAC and Lighting System Performance and Indoor Air Quality of the Oil Sector Complex Building. Progress Report No. 2. KISR 8955.

5. Sebzali, M., A.E. Hajiah, D. Al-Nakib, F. Alghimlas, M. Al-Suwaidan, R. Alasseri. 2008. Provision of Continuous Monitoring & Reporting Services—Indoor & Outdoor Air Quality of Oil Sector Complex Building (Part II). Assessment of HVAC and Lighting System Performance and Indoor Air Quality of the Oil Sector Complex Building. Progress Report No. 3. KISR 9124.

6. Al-Suwaidan M., A.E. Hajiah, D. Al-Nakib, F. Alghimlas, M. Sebzali, R. Alasseri. 2008. Provision of Continuous Monitoring & Reporting Services—Indoor & Outdoor Air Quality of Oil Sector Complex Building (Part II). Assessment of HVAC and Lighting System Performance and Indoor Air Quality of the Oil Sector Complex Building. Progress Report No. 4. KISR 9239.

7. Akbari, H., M.G. Morsi and N.S. Al-Baharna. 1996, Electricity saving potentials in the residential sector of Bahrain, Berkely National Laboratory, Energy and Environment division, LBL-38677, Berkely, USA.

8. Central Bank of Kuwait (CBK). http://www.cbk.gov.kw/WWW/index.html

Chapter 17

Intelligent Lighting: Meeting Your Project's Sustainable Goals

Meg Smith

OVERVIEW

Intelligent lighting is a lighting system that utilizes bi-directional communication between control and luminaire, with the intelligence residing in each uniquely addressed ballast.

The intelligence that resides in the ballast is based on a protocol called DALI (digital addressable lighting interface), an internationally recognized open protocol dedicated purely to lighting control, created to allow interoperability between dimmable lighting products, sensors and devices from different manufacturers. As a result of the intelligence residing in the ballast, control communication is independent of the power circuits therefore eliminating the need for dimming and/ or relay panels and the home runs from the load to these panels.

Essential "intelligent" characteristics of a DALI installation are: (1) individual addressable control and monitoring of lighting systems; (2) precise and repeatable digital dimming; (3) design flexibility and ease of installation; and (4) OPC server for BMS clients.

Intelligent lighting supports the goal of sustainable lighting design: To provide the best quality visual environment with the least impact on the natural environment. Intelligent lighting provides design flexibility which helps the design professional maximize energy savings while optimizing lighting system performance and enhancing the visual environment.

This chapter explores the elements of sustainable lighting design and the general principals of energy smart application guidelines. Case studies of intelligent lighting projects and their sustainable design goals are presented.

How Intelligent Systems with DALI Work

- Power supply sends 16 VDC digital signals over a communication loop.
- Ballasts have integral processors that can accept and respond to the digital signal.
- 256 digital step full range dimming ballast (1-100%), with memory for unique address, and is programmable for (16) scenes & (16) groups.
- Low voltage interfaces available.
- Metal halide ballasts are available, though a DALI relay recommended.
- Incandescent DALI relay control.
- Input device can broadcast group/scene messages to the ballast, interfaces, and relays.
- DALI ballast and devices have 100% interoperability.

Through the tools of intelligent lighting the designer can provide a dynamic lighting system, responsive to the architecture and the users of the space.

Intelligent lighting can provide 5 levels of lighting control:

1. User initiated control commands
2. Automated sensor initiated controls
3. Sensor plus enhanced user interface including sensor enabled and disable group/scene control
4. Centralized building wide lighting system management control
5. Enterprise wide multi system controls

The technology is scalable to the needs of your project and what the project budget will support. Users choose the level of technology and the cost that are appropriate.

Types of Intelligence:

- Stand alone intelligence: dual sensor, smart switch, and individual control. Simple, cost effective, out of the box "plug and play" without commissioning. Pre DALI protocol, DSI "normally off" technology, no constant data. No network options.
- Intelligence based on DALI ballasts and an "extended DALI" protocol (proprietary) addressable sensors. These provide non-

loop specific, flexible, smart automated control.

- Intelligence based proprietary digital networks, These systems use 0-10v ballast technology without bi-directional communication with a proprietary digital interface. These systems are not based on an internationally recognized protocol and do not offer precise and predictable digital dimming or interoperability of devices. They do provide administrative and limited personal control networks.

- DALI network provides personal, automated, administrative, BMS and enterprise-wide control. DALI is the only personal lighting system that provides personal lighting control using the building Intranet. This is accomplished by having communication hubs with a fixed IP address. This means every fixture on the hub can be individually addressed but you only need a few fixed IP addresses. The users are accommodated through floating IP addresses by recognizing the computer name of the user. Most facilities use floating IP addresses. This matters in a large installation with many users. Flexible powerful administrative software for loop specific control DALI electric lighting and day lighting shade control devices.

Wireless and power line carrier intelligent networks are emerging technologies, possibly based on the DALI protocol. The savings and flexibility of a self-repairing wireless network combined with the simplicity of DALI are exciting.

SUSTAINABLE LIGHTING DESIGN GOALS AND APPLICATION GUIDELINES FOR ACHIEVING THOSE GOALS

The IESNA and the IALD define sustainable lighting design as "...meeting the qualitative needs of the visual environment with the least impact on the natural environment. Visually effective and appealing, high quality lighting provides the greatest environmental and economic value."

The greatest single impact lighting has on the natural environment is through its consumption of electricity generated by fossil fuel generation plants. These plants produce carbon dioxide pollution, as well as sulphur dioxide and nitrous oxide that contribute to acid rain.

These plants also release mercury into the atmosphere, contributing to the bio-accumulation of toxic methyl mercury in fish stocks around the globe. Sustainable design practice that incorporates energy efficient lighting solutions, resulting in power demand reduction, in combination with regulations restricting these toxic emissions, can significantly reduce these negative environmental impacts of electricity generation. For this reason, much of this discussion regarding sustainable lighting design does concentrate on energy management. This is not to minimize the importance of other factors which are also considered, including light pollution, materials use and those that focus on the response of the space to its users' ongoing needs.

The Elements of Sustainable Lighting Design:
- Optimizing the use of day lighting
- Minimizing the use of energy through integrated design and effective controls
- Avoiding skyward illumination to reduce light pollution
- Specifying environmentally preferable materials and equipment
- Ensuring system flexibility, maintainability, and durability
- Providing for proper commissioning

Optimizing the Use of Day Lighting: "The Building As Luminaire"
For sustainable designs, daylight should be considered as the primary light source whenever it is available. Lighting design for the interior of a building begins with the shape, orientation and fenestration of the building form. The building itself can make a critical difference in the overall sustainability of a lighting solution.

Daylight should be fully integrated with interior lighting and lighting control systems. In addition to enabling significant energy savings for lighting, daylight can provide a comfortable environment which affects users at both the psychological and physiological levels. Daylight provides optimal color rendering and enhances visual performance.

Day lighting control strategies must be discussed and included during the development of a schematic design, since potential problems such as glare and heat gain will be difficult to correct after the fact. Advance consideration of inter-related design and control elements can result in a day lighting design that balances glare control with energy savings, thermal comfort and performance.

Minimizing the Use of Energy through
Integrated Design and Effective Controls

Energy Conservation: An effective sustainable lighting strategy integrates daylight design and architectural elements with electrical lighting and lighting controls to provide illumination. If this system uses the most efficient sources and luminaires currently available and these systems are designed to deliver illumination levels appropriate to the various visual tasks, then very large energy savings can be anticipated in both energy usage for illumination and energy for space cooling.[1]

For a day lighting strategy to be effective, and for energy savings to be maximized, the electrical lighting systems must be automatically turned off or dimmed as much as possible whenever day lighting in a space is adequate for the current task. In addition, relying on automated control systems for electric lighting (dimmers) and, if possible, exterior shading (automated blinds) proved to be one of the most cost-effective strategies to reduce energy consumption significantly.[2]

In addition to day lighting considerations, there are some basic "Energy Smart " lighting application guidelines for minimizing the use of energy by the electric lighting system.

We should provide lighting only where and when we need it. Utilize a task ambient approach, reducing ambient light levels and tuning the task lighting to the light level appropriate to the task. (see ANSI/IESNA RP-1) Using intelligent lighting systems these levels can be modulated according to time of day as well. For instance, less light is required to perform the same visual tasks at night when the eye adapts to lower ambient levels than the same task performed during the day.

It is critical that vertical surfaces are illuminated. Bright walls and surfaces provide the eye an opportunity to focus away from the immediate task, reducing eye fatigue. Bright walls and surfaces "balance" daylight contribution and create the perception of a brighter more pleasant space which increases user satisfaction. Interior finishes can have a great impact on the ability of lighting equipment to deliver light to various task surfaces and to create the type of lighting scenarios encouraged by the most recent findings on the quality of the visual environment.

Use Reflective Finishes

Ceiling materials are available which are at least 80% reflective. Private office walls and open office partitions should be at least 60-70% reflective.

Dark colors should be used sparingly for local accents.

Large areas of dark surfaces (floor, ceilings, wall panels) absorb more light than light colors and can affect the performance of the lighting system by approximately 20% of its output.

In addition to the advantages of higher reflectance finishes, the designer should be aware of the potential for glare with interior specular and glossy finishes, particularly in daylit spaces.

Use the Most Efficient Lamp/Ballast/Optical Combination Appropriate to the Lighting Task

One cost effective strategy to optimize system efficiency is the use of "high performance T8" lamp/ballast pairings. These pairs can be selected in order to "tune" the ballast factor and target light level to use the least amount of power. The ballast factors for general lighting applications run from .60 to 2.0. Ballast factor, also known as relative light output, is multiplied by the lamp lumens to give you a system lumen number. To calculate the lumens per system watt, determine the ballast input wattage paired with the lamp of your choice from the ballast manufacturers' catalog or data sheet. Some lighting manufacturers "do the math" for you providing you with suggested lamp/ballast combinations and the photometrics for those combinations used in their fixtures. These high-performance T8 systems are of course dependent on efficiency of the optical system to deliver that light. These systems are static, so they only impact power allotment, not the actual energy usage (power over time) and are not appropriate for dimmed daylight harvesting. Efficiency in lighting fixtures is evaluated in terms of a ratio of lumens (the amount of light delivered) per watt consumed. Wattage is the unit of measurement for power, so this ratio is a description of the power required to produce an amount of light. It does not address actual energy usage. This ratio is a convenient short-hand method of evaluating various light fixture options. Care should be taken to also consider the light distribution, spacing ratios and shielding for visual comfort.

A second metric for describing efficiency is the LER or luminaire efficiency rating. This addresses the efficiency of the luminaire within the context of its application. One luminaire/light source combination may use a lower efficacy source or a less efficient optic and still be more effective than another at delivering lumens to the task or distributing light where it is needed. For instance, an LED under counter light may deliver exactly the amount of light needed to the counter work surface for much less energy than a similar luminaire with a halogen source, even though the halogen source itself may have a higher source efficacy. A direct/indirect optic may have a lower efficiency ratio than another, but it may be more effective at distributing the lighting uniformly over the ceiling plane, allowing for a wider spacing ratio requiring fewer fixtures and therefore reducing power density.

Integrating Controls can Affect the Energy Efficiency of a Lighting System by Impacting the Hours of Use

Integrated intelligent controls layer automated controls with user-initiated controls and scheduled events. Intelligent lighting controls enable the lighting system to deliver the right amount of light to the right place at the right time. Energy use can be reduced by significant amounts, easily between 20-50%.

The goal of sustainable lighting design is to provide the best quality visual environment with the least impact on the natural environment. Intelligent lighting can assist in achieving that goal by providing design flexibility. This flexibility assists the design professional to maximize energy savings while optimizing the performance of the lighting design and enhancing the visual environment.

A Designer can Achieve Aggressive Energy Savings by Integrating Multiple Controls into the Space

Intelligent lighting controls organize and prioritize layered control strategies. Three types of controls that can work together to save energy are day lighting, occupancy sensing and individual user controls. Task tuning can be enhanced through scheduling.

The results from this field study show the combined energy savings from occupancy, day light and individual user control to be 65%-75%. The lower number reflects interior and second daylight zone and the 75% is the data for the perimeter zone.

By analyzing the polling data, the effect of each control on energy savings is deduced: occupancy alone ranged from 49¬55%, light sensing from 23-35%, and individual user control from 12-19%. While each project is unique and its lighting control design should be also, studying this chart may assist in making cost effective control choices.

The data demonstrate the effectiveness of layering energy man-

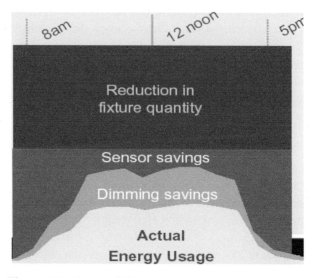

Figure 17-1. Layered Energy Management Strategies

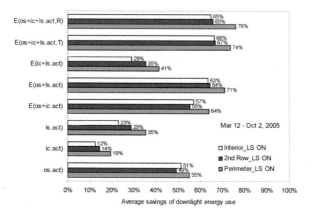

Figure 17-2. Energy Results from a Controls Field Study Done by Canadian National Research Council in Conjunction with Bc Hydro

agement strategies. Particularly interesting is the energy savings tendency found when users are given control of their lighting. Earlier studies had demonstrated its effects on user comfort, satisfaction and productivity. Now we know we can provide these benefits while saving energy.

Ensuring System Flexibility, Maintainability, and Durability

Sustainable projects are incorporating control systems in order to meet the building's demands of energy management and the user's needs for system performance. These lighting systems should also be designed to ensure system flexibility and maintainability in order to extend the useful life of the facility and to better serve the user's changing needs. The lighting design should anticipate change and the systems should be durable enough to survive system reconfiguration and reuse in future remodeling.

For the facilities manager (who is often not an electrician or systems integrator), simplicity is also a favored characteristic of the lighting system. The designer should take into consideration the: complexity of the system, number of components, ease of system reconfiguration, facility monitoring and reporting, maintenance diagnostics, availability and ease of replacement of components. A partially functioning or abandoned control system is neither saving energy nor contributing to the accomplishment of the building's sustainable goals. The components of a non-functioning system are likely to end up in the waste stream prematurely!

The sustainable lighting principals outlined above translate into credit areas under Leadership in Energy and Environmental Design (LEED®) Green Building Rating System guidelines.

LEED® is a voluntary rating system that evaluates a project within 5 areas. They are: Energy and Atmosphere, Indoor Environmental Quality, Materials and Resources, Sustainable Site and Water Usage. Within these criteria called credit areas, points can be earned toward a level of certification.

LEED® programs have been developed for a variety of construction and building types and phase of a building lifecycle. Specific LEED® programs include: New Commercial Construction and Major Renovation, Exiting Building Operations and Maintenance, Commercial Interiors, Core and Shell Development projects, Homes, Neighborhood Development, Guidelines for Multiple Buildings and On-Campus Building Projects, LEED® for Schools and LEED® for Retail. Under development are LEED® for Healthcare and LEED® for Labs.

Each program has a slightly different set of point requirements for achieving LEED® certification. Emphasis is placed on different credit areas depending on the type of construction or the application.

Points for Leed Certification		
	LEED-NC	**LEED-CI**
Certified	26-32	21-26
Silver	33-38	27-31
Gold	39-51	32-41
Platinum	52-69	42-57

Figure 17-3. Points for LEED Certification for New Construction and Commercial Interior

As of June 26, 2007, all LEED-certified projects are required to achieve at least two "Optimize Energy Performance" points within LEED, which will improve the energy performance of all LEED certified green buildings by 14% for new construction and 7% for existing buildings.

Although lighting is excluded from being a requirement for materials use credits, they can be used to contribute to earning these points. Lighting specifiers should consider: lower wattage/low mercury sources, Nickel metal hydride batteries for exit signage, and integrated sensors for component and wiring reduction.

The LEED® Existing Building standard does address disposal and recycling issues. In order to encourage process improvements, it is planned that all LEED for New Construction and Core and Shell buildings that reach certification will automatically (at no cost) be registered for LEED for Existing Buildings. Existing Buildings are enrolled in an on-going certification program. For these projects maintenance and commissioning and documentation and verification of systems importance are critical components of their certification.

In addition to the new 2-point energy optimization requirement, the USGBC has new CO_2 emissions reduction goals. All new commercial LEED projects will be required to reduce CO_2 emissions by 50% of current emission levels in future versions of the programs There are plans for the implementation of a carbon dioxide off set program.

LEED Credits Related to Lighting		
Credit	New Construction (NC 2.2)	
	Points	Action
Minimum Energy Requirement		ASHRAE 90.1-2004
Minimizing Energy	1-10	Reduce total energy consumption 10.5-42% (modeled)
Daylight	1	Provide 25 FC from daylight to 75% of space
Views	1	Provide window view to 90% of occupants
Controls	1	Provide individually controllable lighting to 90% of occupants and shared spaces
Light Pollution	1	Reduce exterior lighting power beyond Standard 90.1 Control impact of interior lighting, sky glow and light trespass

Figure 17-4. LEED Credits Related to Lighting

ASHRAE Standard Project Committee 189 (SPC 189) Standard for the Design of High-performance, Green Buildings Except Low-Rise Residential Buildings*

At the time this chapter was written, this document was in the public review stage. Although it is premature to report conclusively on specific requirements, Standard 189 is too important a document to let go by unmentioned. This document represents the first cooperative effort between, ASHRAE, IESNA and USGBC and it will become

*The standard has been adopted.

LEED Credits Related to Lighting		
Credit	**Commercial Interiors (CI-2.0)**	
	Points	**Action**
Minimum Energy Requirement		ASHRAE 90.1-2004
Minimizing Energy	1-3	Reduce lighting power density 15-35%
Daylight	1-2	Provide 25 FC from daylight to 75% or 90%of space
Views	1	Provide window view to 90% of occupants
Controls	1	Provide individually controllable lighting to 90% of occupants and shared spaces
Light Pollution	0.5	Meet IESNA RP-33 Control impact of interior lighting, sky glow, and light trespass
Enhanced Metering	2	Lighting systems and controls (and all other systems)
Daylight Controls	1	Daylight- responsive controls within 15' of windows and under skylights
Enhanced Commissioning	1	Commissioning Authority and integrated manual

Figure 17-5. LEED Credits Related to Lighting for CI 2.0

the benchmark for all sustainable green buildings in the United States because it is being developed for inclusion into building codes. It is the first such green building standard in the United States.

Standard 189 is remarkable in a number of aspects. It is a standard that addresses energy usage as well as power densities. The goal is to achieve a minimum of 30 percent reduction in energy cost (and carbon dioxide equivalent) over that in ANSI/ASHRAE/IESNA Standard 90.1-2007. Lighting power densities are 90% of current ASHRAE 90.1 levels. There is a requirement for automatic load limiting or shifting for at least 5% of the electrical service load.

The most significant lighting issues addressed are the requirements for automated control systems and the integration day lighting and daylight harvesting.

Many of the requirements include automated (sensor input) controls. In most spaces, the size of the occupancy control zone has been reduced to 1000 square feet, 250 square feet in private offices. All class rooms, lecture, training or vocational rooms must have occupancy control. Occupancy control shall be manual-on or auto-on with programmable reduced light levels and occupancy response combined with manual-on for brighter setting. Occupancy will work with day lighting controls in daylight areas. Daylit areas will have full range dimming or multi level switched controls. Vacancy in hotel and motel hallways, commercial and industrial storage stack areas and library stack areas will require a sensor response with multi-level switching or a dimming system that reduces lighting power a minimum of 50%.

All daylit zones shall be provided with controls that automatically reduce lighting power in response to available daylight by either: a combination of dimming ballasts and daylight-sensing automatic controls, which are capable of dimming the lights continuously, or a combination of stepped switching and daylight-sensing automatic controls which are capable of incrementally reducing the light level in steps automatically and turning the lights off automatically. An additional credit is proposed for zones controlled separately beyond the 15' perimeter zone.

Sensor control is extended to outdoor lighting. Areas with two or more luminaires must provide for automatic time clock switching of 50% of the lighting when not needed and/or dimming to be provided which will control within the range of 50-80% of the light output.

Much attention is spent on promoting the proper use of day

lighting, including a new provision for top lighting in buildings of 3 stories or less over grade and over 20,000 sq feet with 12' ceilings and power densities over .5 watts/sq ft. The requirements for both top lighting and side lighting include minimum fenestration areas, glazing standards to assure low heat gain, shading to prevent direct sun (glare) and interior surfaces visible light reflectance requirements. The reflectance levels are greater than or equal to 80% for ceilings and 70% for partitions higher than 60 inches. Classrooms and offices shall have 30 foot candles of usable illumination provided by day lighting within the 15 perimeter area from the façade.

Emphasis is placed on ongoing performance of all systems. Lighting and shading system controls must be commissioned prior to occupancy. All automatic day lighting controls, manual day lighting controls, occupancy sensing devices, and automatic shut-off controls must have acceptance testing.

THE ENERGY POLICY ACT OF 2005 (EPACT 2005)

The Energy Policy Act of 2005 (EPAct 2005) provides tax incentives for buildings whose energy performance reaches or exceeds 50 percent above ASHRAE 90.1-2001. Commercial buildings and/or equipment entering service between January 1, 2006 and December 31, 2008 can realize a tax deduction of $1.80 per square foot by using 50 percent less energy with respect to lighting, HVAC, envelope and hot water systems. Significant savings can also be realized by addressing individual elements.

Lighting upgrades provide significant energy savings. EPAct specifically allows it to be considered as a separate system, offering a deduction of up to $0.60 per square foot for lighting alone.

EPAct 2005's prescriptive method or "standard" requires reduction of LPD (lighting Power densities) below AHRAE 90.1 as a prerequisite. That LPD requirement refers to connected load, which must be accomplished through high performance luminaires and energy smart applications. EPAct 2005 also requires bi-level switching, which, under EPAct 2005 guidelines, can be accomplished through dimming as well as switching.

Dimming reduces energy costs more than bi-level switching. Those ongoing savings go to the owner of the building/energy costs, continuing to payback after the one time tax credit has been collected.

The energy cost budget method allows building owners who want to utilized dimming strategies such as load shedding, daylight harvesting or adaptive compensation which do not reduce lighting power density as specified in the Standard, to qualify for tax credits. The energy cost budget method, requires computer modeling.

Below are brief descriptions of four successful projects, aided by the technology and application strategies of intelligent lighting.

EPAct 2005 Tax Incentives for Lighting		
Reduction of LPD below Std 90.1-2001	LPD[1] Watts per SF	Eligible Tax Deduction per SF[2]
25%	.98	$0.30
26%	.96	$0.32
27%	.95	$0.34
28%	.94	$0.36
29%	92	$0.38
30%	.91	$0.40
31%	.90	$0.42
32%	.88	$0.44
33%	.87	$0.46
34%	.86	$0.48
35%	.85	$0.50
36%	.83	$0. 52
37%	.82	$0. 54
38%	.81	$0.56
39%	.79	$0.58
40%	.78	$0.60
> 40%	.78	$0.60

Figure 17-6. Epact 2005 Tax Incentives for Lighting Upgrades Intelligent Lighting: The Smart Way to go Green: Specific Case Study Discussions

MAXIMIZING ENERGY SAVINGS:
PROJECT: TENNESSEE VALLEY AUTHORITY
BLUE RIDGE COC PILOT STUDY

The TVA is the nation's largest public power company. TVA provides power to about 8.7 million residents of the Tennessee Valley. In response to executive orders and EPAct 2005 calling for federal facility energy savings, the TVA was looking to maximize energy savings in its own facilities in seven states as well as provide a model of conservation to its 8.7 million customers.

Other priorities were also important to this project. There was a desire to realize the day lighting design of the TVA's Chattanooga headquarters' architecture. The original scheme had been abandoned during its construction due to cost and maintenance concerns. The TVA had a need to create a universal solution that would fit all new and retrofit projects throughout its seven state real estate holdings. There was a real concern for the ability to satisfy the visual needs of a diverse and aging workforce.

It was decided that they would develop a universal furniture based lighting automated system; adaptable to various architectural conditions, work group configurations and user requirements.

The Lighting System Included Four Levels of Control
1. Enterprise network capabilities for remote energy use and facility monitoring.
2. Building-wide network for scheduled control, task tuning, and peak load shedding.
3. Automated (sensor based) input for daylight harvesting and occupancy response
4. Individual user control for maximum flexibility, adaptability and to provide an ergonomically responsive lighting system which also encourages energy savings.

The luminaire was selected to support the visual requirements of the aging eye. It provides superior glare control for "task" portion, a distribution for good indirect lighting uniformity within task area, adequate light levels, consistent with IES RP28-98 (50fc task, average 44 fc under bin) while providing superior optical control to create low glare/high contrast task areas. (30 fc ambient)

Technology Solutions
- The system is a network using the existing LAN and a lighting server with windim@net software for administrative and individual control.
- The DALI fixture based sensors remove the automated controls from the ceiling and the power circuit

Energy Savings Data
- Power density (connected load in watts per square foot) reduced by 52% or 1.1 watt per square foot
- Layered energy management strategies reduce average energy usage an additional 53% below connected load depending on time of day and user requirements.
- The effective power density at peak load is .7 watts per square foot and the average of the project is .565 watts per square foot.

This work station specific mounting provides a TVA system wide "universal" lighting solution which: incorporates best practices, accommodates diverse work force including the needs of the aging eye, provides ergonomic lighting system to enhance workplace productivity and saves energy.

Figure 17-7. TVA Pilot Study Workstation Specific Lighting

OPTIMIZING SYSTEMS:
PROJECT: QUEENS UNIVERSITY, ONTARIO CANADA

This example shows where solar power, daylight harvesting and intelligent lighting go to the head of the class. The system performs and teaches! The Intelligent Learning Centre (ILC) uses its website to monitor building systems. See: http://appsci.queensu.ca/ilc/live-building/lighting/index.php to view real time monitoring of the lighting system!

In the first project, intelligent lighting provides design flexibility to maximize energy savings while optimizing system performance and enhancing the visual environment.

The ILC is a green building with a pedagogical function, housing six faculties of engineering in this 100,000-square-foot three-story building, arranged around a central atrium. This visual connectivity throughout the facility is intentional. The intent was that the building should educate and increase awareness of environmental and sustainable issues, such that the students may take the valuable lessons learned with them to their careers in professional practice.

The atrium also provides additional clerestory lighting through a glazed sawtooth roof section. To maximize the effect, the soffit of the ceiling of this diagonal skylight is coated with highly reflective material to bounce the light into the terraced interior plazas. The lighting system was designed to achieve an integrated, sequenced lighting concept approach to maximize day lighting. Offices are lit with a combination of daylighting and light shelves.

Careful computer modeling of the interaction of day lighting and artificial lighting were carried out. The ILC integrated the day lighting with the intelligent lighting controls.

Various energy-saving systems are employed at the ILC. High-efficiency lighting systems include an intelligent lighting network including centralized and local controls, photocells and occupancy sensors that have been installed to optimize efficiency.

All offices have optically-efficient luminaries which include: integral individual occupancy sensors with an IR port for personal dimming in offices with glass walls and programmable to manual-on, auto-off for offices and class rooms which had wall controllers as well.

This was the first installation with a DALI group/scene interface for sensor and wall box inputs. More than 60 of these interfaces, de-

Figure 17-8.
Queen's University Intelligent Learning Center Day Light Intergration

signed for this project were used in lecture halls, corridors and meeting rooms.

The network sets maximum light levels to 50 footcandles. Occupants are able to dim the lights from their computers, but not able to exceed the maximum.

The reviews are all positive. BREEAM's Green Leaf Eco-Rating Program gave the ILC a "Four Green Leafs" rating out of a possible five. The building was found to be 25.9% more efficient than the MNECB (similar to our Energy Star program) reference building. Perhaps the most important review came at a post occupancy meeting last year, when the project manager and now buildings systems manger commented the intelligent lighting system "was working exactly as planned."

INTELLIGENT LIGHTING CAN HELP YOUR CLIENT ACHIEVE THEIR SUSTAINABLE GOALS PROJECT: HARVARD SCHOOL OF PUBLIC HEALTH

Landmark Center embraced sustainable design while promoting health, comfort and increased staff productivity. It attained LEED™ Silver Certified rating, for this fit-out in an old building.

In keeping with its mission and in order to provide a living laboratory, "devising new strategies for a healthier environment, a safer workplace, and fewer injuries," Harvard School of Public Health leased 40,000 sq. ft. of commercial office space in the historic Landmark Building, which was originally a Sears-Roebuck warehouse.

A major feature of this sustainable renovation is the low energy lighting system with its efficiency and effectiveness optimized by the first DALI Lighting Control system installed in North America.

The Harvard School of Public Health documented a 40% reduction in lighting demand. Lighting power was reduced through the use of optically-efficient T5HO luminaires, occupancy programming and sensor control as well as individual user control, reducing the size of control zones and thereby increasing comfort and productivity as well as maximizing energy savings. Estimated savings in energy coupled with productivity gains resulted in a ten-month payback for green features. The building also achieved EPA/DOE Energy Star recognition award for building energy efficiency.

"Incorporating USGBC LEED standards into University practices makes sense from all angles: financial, environmental, and human health."—Daniel Beaudoin, CEM, Manager of Operations, Energy & Utilities.

The innovation credit for the LEED TM certification states: "The building is used in graduate research; it is used in the education of staff, students, and faculty; it has been used to assist in the market transformation towards a more sustainable building industry through leading tours for industry."

HOW INTELLIGENT LIGHTING
HELPS YOUR CLIENTS MEET THEIR BUSINESS GOALS.
PROJECT: ASSOCIATED PRESS (AP) WORLD HEADQUARTERS,
NEW YORK, NEW YORK

Intelligent lighting satisfies competing visual tasks in the 84,000 square foot open floor plan with 1700 workstations while saving energy and money.

The performance criteria drove this project. AP needed to provide a high-performance visual environment that was flexible enough to survive a compressed design/construction schedule as well as post-occupancy space reconfiguration and adaptable enough to satisfy competing visual needs within an open floor plan.

Combining full range dimming task tuning with time scheduling made the system both responsive and economical. Non-circuit base control allows post occupancy reconfiguration.

A conference center above the news floor was linked to the same network. At no additional cost, the network provided a "bonus" partitioned space lighting control system (replacing the expensive conventional wall box system originally specified by the architect), which easily handles the reconfiguration requirements of the conference center.

At the start of the project, the owner requested a lighting solution that would make the employees "stop complaining" about the lighting. A month after commissioning the project, Associated Press

Figure 17-9. Associated Press World Headquarters, New York City. Balancing Competing Visual Needs

reconfigured the news floor to cover the Republican National Convention and again to expand the sports department for the 2004 Summer Olympics. In November, the conference center was transformed into a reporters' bullpen to cover the national election. Robert Heizler, lead architect reported, "no complaints about the lighting!"

SUMMARY

The features of intelligent lighting control systems that contribute to the success of a sustainable design project are:
- Flexible daylight responsive lighting control solutions
- Enhanced task tuning, scheduling and demand response
- Control and prioritize a variety of inputs: day lighting, occupancy, AV, BMS
- Personal dimming control
- Scalable technology provides cost effective design solutions adaptable to the users' needs
- Smaller control zones with reduced wiring, materials and components.
- 30-60% lighting energy savings beyond ASHRAE 90.1 2004
- Documentation & verification

References
1. See e.g.: Krarti, 2005, "A simplified method to estimate energy savings of artificial lighting use from daylighting," Building and Environment, vol. 40; Jenkins and Newborough, 2007, "An approach for estimating the carbon emissions associated with office lighting with a daylight contribution," Applied Energy; Franzetti et al., 2004, "Influence of the coupling between daylight and artificial lighting on thermal loads in office buildings," Energy and Buildings vol. 36.
2. Leslie et al., 2005, "The potential of simplified concepts for daylight harvesting," Lighting Research and Technology, vol. 37.

Chapter 18

New Lighting Technologies Demonstrated at Defense Commissaries

Steven Parker, PE, CEM
Joseph Konrade
E. Carroll Shepherd III, CEM

OVERVIEW

New and emerging lighting technologies, such as LEDs (light-emitting diodes), can improve lighting quality while reducing maintenance and energy costs. The Defense Commissary Agency (DeCA), with support from the Department of Energy's (DOE) Federal Energy Management Program (FEMP) and the Pacific Northwest National Laboratory (operated for DOE by Battelle Memorial Institute), demonstrated the use of LED lighting in a large freezer storage room and fiber optic lighting in a series of vertical reach-in display freezer cases at the Fort George G. Meade Commissary. The LEDs reduced lighting energy use by 85% and reduced maintenance requirements. The fiber optic lighting system reduced lighting energy use by 56%.

BACKGROUND

The Defense Commissary Agency (DeCA) operates a worldwide chain of commissaries providing groceries to military personnel, retirees and their families. Authorized patrons save an average of more than 30 percent on their purchases compared to commercial prices, which is a valued part of military pay and benefits. For these reasons, keeping costs low is paramount to the Defense Commissary Agency.

Commissaries, like most food sales facilities, are some of the most energy intensive buildings. Unlike most buildings, the largest energy consumer is usually the refrigeration system, which can account for half of the total annual energy consumption. The second largest energy

257

consumer is usually the lighting system. Together, these two account for significant annual operating costs. Reducing these energy costs is a major concern to the Defense Commissary Agency.

Commissaries have extended customer business hours compared to most buildings. During the day, patrons keep the front of the building busy. Deliveries keep the back of the building busy as trucks continuously arrive and products are unloaded and placed in interim storage. After customer hours, shelves are restocked with the products from storage. The Fort George G. Meade* Commissary, considered one of the busiest commissaries, is no exception.

INTRODUCTION

The most visible energy consumers in the commissary are the refrigerated reach-in display cases. Lining several aisles in the commissary, vertical reach-in display cases store frozen and refrigerated foods and are directly accessible by patrons. Food products need to be kept cold, while being well illuminated and highly visible.

To support the high business volume, the Fort George G. Meade Commissary has several large freezer storage rooms located in the rear of the commissary facility. Different freezers are kept at different temperatures depending on the products stored; some as low as -20°F. The storage rooms are very busy and very energy intensive. In addition to the refrigeration load, each room is lit to provide adequate illumination.

Conventional lighting and refrigeration systems typically work against each other. Lighting systems generate heat, which the refrigeration system needs to remove. In addition, lower temperatures typically reduce the efficacy of lighting systems. Thus, more power is required to generate the desired illumination, which in turn, increases the load on the refrigeration system.

Investigating ways to reduce energy consumption and costs, the Defense Commissary Agency and the Fort George G. Meade Commissary sought to demonstrate new lighting technologies in the re-

---

*Fort George G. Meade, located a few miles south of Baltimore, Maryland, is home to the fourth largest workforce of Army installations in the continental United States. Fort Meade's workforce has approximately 39,000 members, composed of military, civilian and contractor personnel. The local population is estimated to be 109,000, in addition to thousands of daily visitors.

frigerated spaces. In the back of the commissary, between the loading docks and the main customer sales area, LEDs (light-emitting diodes) were demonstrated as a replacement for the old incandescent lighting system in a large drive-in freezer storage room. In the main customer sales area, a new fiber optic lighting technology was demonstrated in a series of vertical reach-in freezer display cases.

LED LIGHTING SYSTEM

The 35- by 47-ft freezer storage room is kept at -20°F and is designed to accommodate large pallets of frozen food, including ice cream products. The old lighting system consisted of 36, 100-watt gel-coated incandescent lamps in globe-type enclosed fixtures mounted on the ceiling; 13-1/2 ft above the floor (see Figure 18-1). While there is a light switch located outside the main door from the inner loading dock area, the freezer storage room has heavy traffic day and night, and the lights are rarely turned off.

To reduce lighting and refrigeration energy consumption, while also reducing maintenance requirements, the old incandescent lights were replaced with 36, 15-watt white LED fixtures from Energy Focus, Inc. (see Figure 18-2).

The LED globe lights* offer several advantages over the incandescent lamps.

- The new LED lighting system provides over 10 footcandles of illumination on the floor, an improvement over the incandescent lamps, even when all the lamps were working. Lighting power is reduced to 540 watts from 3600 watts, a reduction of 85%.

- The new LEDs deliver a different correlated color temperature (CCT) compared to the old incandescent lamps; 6500K for the new LEDs compared to 2600K for the incandescent lamps. The scotopically enhanced color from the new LEDs provides an improved perception of overall brightness.

- LEDs are inherently directional light sources. In this application,

---

*Armada LED small globe, 15 watt, white, 6500K, 750 rated lumens by Energy Focus, Inc., formerly Fiberstars, Inc. (www.energyfocusinc.com)

Figure 18-1. Freezer Storage Room
with Incandescent Lamps

Figure 18-2. Freezer Storage Room
with New Led Globe Lights

it means the lamp-fixture LED system is more effective at getting useful light to the task. To improve light distribution, the new LED globe enclosures are frosted.

• Normally, heat management is a design issue with LED fixtures. In this application, the cold storage environment actually improves the efficacy of the LED light source. Unlike conventional lamps, the light output of LEDs improves in cold climates. At -20°F, the light (lumen) output of the LED light is about 18% greater than at normal room temperatures.

Another source of energy savings is through the refrigeration system. All energy consumed by the lighting system becomes heat load for the refrigeration system to remove. Therefore, less power to the lighting system means reduced load on the refrigeration system. To see if the impact on the refrigeration system was measurable, the lights and refrigeration system were sub-metered for 3 months as part of this demonstration activity supported by FEMP. Unfortunately, the routine daily variance in refrigeration compressor power was too great compared to the lighting power reduction to see a statistically significant difference.

FIBER OPTIC LIGHTING SYSTEM

The reach-in freezer cases are kept between –5°F to –10°F and are used to display a variety of frozen foods from quick preparation meals to ice cream. The vertical reach-in display cases used in the demonstration line the entire length of aisles 9 and 10, and include a total of 79 access doors. The previous lighting system consisted of 87, F40-T8 (60-inch) fluorescent lamps with customized electronic rapid-start ballasts.* The operation of the lights is regulated by a digital control system. Operating hours vary by day of the week but average 94-1/2 hours per week.

---

*Ardco EcoTronic™ electronic rapid start ballast, model number 2/40ET-T/120V/LTHP, catalog number C15724 P8, 2-lamps per ballast, low temperature, 120-volt, 1.02-amps, power factor is 0.96 or above. Rated input power is approximately 117.5 watts (per ballast).

To reduce energy consumption, the old fluorescent lighting technology was replaced with a new system that uses fiber optics, shown in Figure 18-3. The fiber optic lighting system uses a remote source light. The light is channeled into a fiber optic distribution system and emitted into the space by an illuminator. The illuminator uses optics designed to match the application to illuminate the product. In this demonstration, the Energy Focus EFO-ICE™ system was installed. The new lighting technology uses a 70-watt metal halide lamp as a source light. From each source light, up to six fiber-optic cables are used to transfer the light to the vertical reach-in display freezer cases. The source lights are mounted on top of the vertical reach-in freezer cases, as shown in Figure 18-4. The fiber optic cables are routed through the top of the display cases through rigid conduit designed to prevent kinks in the fiber (if the fiber bends too sharply, light transmission can be significantly reduced). Inside the display case, the fiber optic cable connects to an illuminator, which distributes the light. An illuminator is installed on each side of a door. In general, one 70-watt metal halide lamp replaces up to three F40-T8 (40-watt) fluorescent lamps. The fiber optic lighting system offers several advantages over the fluorescent lighting system.

•   The new fiber optic lighting system requires less overall power

Figure 18-3. Vertical Reach-in Freezer Display Cases with Fiber Optic Lighting

and energy. Measured lighting power was reduced to 2281 watts from 4968 watts, a reduction of 54%.

• The new metal halide lamp fixtures are located on top of the reach-in freezer cases. This provides easier maintenance access to the lamps and ballasts but also removes the heat source from the refrigerated space.

• There are fewer lamps to maintain. With the previous fluorescent lamps, there were always a number of lamps burned out.

• Fluorescent lamps, while normally an efficient light source, have reduced efficacy in low-temperature environments. Because the metal halide light source is located outside of the refrigerated space, the efficacy remains high.

• The new metal halide lamps deliver a different correlated color temperature (CCT) compared to the standard fluorescent lamps; 6000k for the metal halide compared to 3500k for the fluorescent lamps.

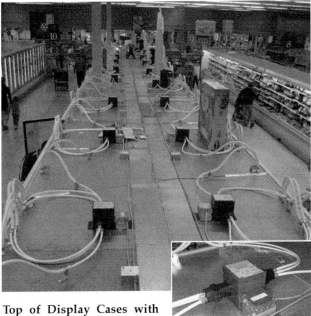

Figure 18-4. Top of Display Cases with Metal Halide Source Lights and Fiber Optic Connections

Another source of energy savings is through the refrigeration system. All energy consumed by the previous fluorescent lighting system became heat load for the refrigerant system to remove. The new metal halide lamps and ballasts are installed outside of the refrigerated space, which should mean less heat load for the refrigeration system to remove. To see if the impact on the refrigeration system was measureable, the lights and refrigeration system were sub-metered as part of a demonstration activity supported by FEMP. The existing system was monitored for 3 months before the fiber optic lighting system was installed in the reach-in freezer display cases. The new lights and refrigeration system were monitored for an additional 2 months after the installation.

While the reduction to the refrigeration load should have been measureable, no reduction in refrigeration power or energy was observed. The lighting heat load on the refrigeration system was reduced by around 5 kW for an average 13-1/2 hours per day. The refrigeration system serving the series of reach-in freezer display cases had a typical load around 26 to 27 kW. Preliminary estimates indicated that peak refrigeration power should have been reduced by about 3 kW (~10%) and energy consumption should have been reduced by about 43 kWh/day (~6.5%).

A number of issues may account for no measurable reduction in the refrigeration energy, such as:

- changes in product turnover or product type affecting the refrigerant load

- changes elsewhere in the system served by the same compressor rack/refrigerant system

- minimum control features or settings on the refrigeration compressor rack.

For comparative purposes, illumination measurements were taken inside a sample of four of the vertical reach-in freezer cases, both with the previous fluorescent lighting system and again after the conversion to the fiber optic lighting system. There are no real standards for illumination levels in this application, and the implication of the results is subjective. Based on the light-level measurements, the illumination on the product was reduced by an average of 50%.

These measurements, however, are with fluorescent lamps with the lumen output in various states of depreciation compared to metal halide lamps in new condition. Compensating for the anticipated lumen depreciation of the metal halide lamps as they age, it is estimated the illumination on the product will be reduced by around 60% compared to fluorescent lamps in similar condition. While the reduction by the numbers appears significant, the subjective appearance of the vertical reach-in freezer display cases is very close, as shown in the side-by-side photo with graphical illustration of the illuminance distribution based on field measurements in Figure 18-5.

OPERATIONS AND MAINTENANCE

The new LED lights are expected to provide over 5 years of useful service even with continuous operation. This contrasts with the old incandescent lamps, which needed to be replaced more than eight times each year. While changing burned out lamps in a -20°F environment is not a pleasant task, significant labor is saved because the frozen foods do not need to be shifted to allow safe access to the overhead fixtures. Because the commissary uses a maintenance contractor for lamp replacement, it means maintenance costs may be reduced as a result of the reduced labor requirement. Not having to replace the burned out incandescent lamps is also a cost savings.

In the case of the fiber optic lighting system, the metal halide lamps are more expensive than the previous fluorescent lamps. While there are notably fewer lamps in use with the fiber optic system, and true lamp life has yet to be verified, it is anticipated that net annual lamp costs will increase a small amount. In the same light, labor requirements for changing the new metal halide lamps are expected to be reduced by about half.

CONCLUSIONS

The new LED lighting system for the freezer storage room at the Fort George G. Meade Commissary is a notable improvement over the older technology. This application takes advantage of the benefits of an emerging lighting technology and serves as an example to others.

Lighting power is reduced by 85%; overall illumination is improved; the load on the refrigeration system is reduced; and maintenance requirements are reduced. As shown in Table 18-1, the LED lighting system provides a good return on investment and allows the Defense Commissary Agency to reduce energy consumption and costs, which assists in achieving the Agency's mission and goals.

The new fiber optic lighting system for the vertical reach-in freezer display cases at the Fort George G. Meade Commissary is an improvement over the fluorescent lighting system. This application takes advantage of the benefits of an emerging lighting technology and serves as an example to others. Lighting power is reduced by 54%; the load on the refrigeration system is reduced; maintenance requirements are reduced; and illumination is maintained. As shown in Table 18-2, the fiber optic lighting system allows the Defense Commissary Agency to reduce energy consumption and costs, which also assists in achieving the Agency's mission and goals.

Table 18-1. Led Demonstration Results

	Before	After	
Average total lighting power	3,600	540	Watts
Operating hours per day	24	24	hours
Energy consumed per year (lights only)	31,363	4,704	kWh
Energy saved per year (lights)		26,659	kWh
Energy saved per year (refrigeration)		indeterminate	
Energy saved per year (total)		26,659	kWh
Energy cost reduction per year (total)[†]		$4,561	
CO_2 reduced per year[‡]		16.1	tons
Installed Cost		$14,400	
Simple Payback		3.2	years

[†] Assumes electricity cost = $0.1711/kWh. (Reference: Fort Meade FY06 Energy Management Report)
[‡] Assumes 1.21 lbs CO_2/kWh (Source: EIA 2006)

Figure 18-5. Side-by-side Comparison (Fiber Optic on Left, Fluorescent On Right, Illuminance Scale in Footcandles).

Table 18-2. Fiber Optic Demonstration Results

	Before	After	
Average total lighting power	4968[*]	2281[†]	Watts
Operating hours per day	13.5	13.5	hours
Energy consumed per year (lights)	24,346	11,178	kWh
Energy saved per year (lights)		13,168	kWh
Energy saved per year (refrigeration)		indeterminate	
Energy saved per year (total)		13,168	kWh
Energy cost reduction per year (total)[‡]		$2,253	
CO_2 reduced per year[**]		8.0	tons
Installed cost		$30,000	
Simple payback		13.3	years

[*] Connected power would be higher but a number of lamps are always burned out.

[†] Connected power includes some fluorescent lamps in different display cases.

[‡] Assumes electricity cost = $0.1711/kWh. (Reference: Fort Meade FY06 Energy Management Report)

[**] Assumes 1.21 lbs CO_2/kWh (Source: EIA 2006)

Chapter 19

Innovative Lighting Contest: Small Actions Add Up To Big Savings In Schools

Mike Barancewicz & John Lord

We are the Loudoun County Public Schools Energy Education Team (LCPS). One of our primary jobs is to educate everyone who uses our buildings: faculty, staff, administration, students, and the community about how they can be a part of our energy conservation program by developing and using good energy saving habits.

First, we would like to tell you a little about LCPS. We are one of the fastest growing school districts in the nation. This year we are opening three new elementary schools. With these additions we have a total of 75 schools, over 8000 employees and over 57,000 students. This last fiscal year (FY08), our school district paid over $12.5 million in utility bills. In order to keep utility expenses to a minimum, our school district has a comprehensive energy conservation program.

The reduction of energy expenses in LCPS has long been an ongoing priority since 1993. This is when Energy Education Incorporated and our school system first implemented the LCPS energy conservation program. Since then, with the help of all our facility users, our program has produced in excess of $26 million in cost avoidance. With the rises in utility costs that everyone is experiencing, the energy conservation program has taken on a whole new level of importance. Even with a successful

program, the time was right to do something innovative and different.

With the encouragement of Dr. Hatrick (LCPS Superintendent), a contest was developed. The goals were to create an increased awareness of energy savings habits, foster student involvement in energy savings practices, recognize student achievement, and experience energy savings due to reduced lighting usage. Students were to create a design for a light switch plate sticker that encourages all facility users to turn off the lights in unoccupied areas. All entries were required to be hand drawn, though the text could be computer generated, stenciled, or pressed on.

One of the nicest features of the contest is the fact that Dominion Virginia Power, the Loudoun Education Foundation, and the Northern Virginia Electrical Cooperative (NOVEC) each provided generous donations to help make this contest happen as part of their commitment to community service. The entire cost of the grand prizes was supplied by these contributions. This is truly an example of collaboration between industry, associations, and the public schools.

The contest was very popular among students. All current LCPS students were eligible to enter, with each school holding a preliminary round of judging.

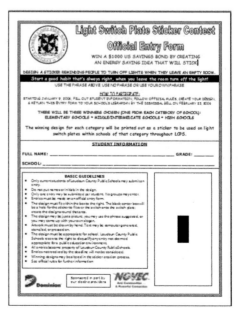

At the school level, the librarians were very helpful in marketing and organizing the contest within their school.

One winning entry from each school was submitted to the LCPS Instructional Materials Center to represent the school for the final judging. The three students who created the grand prize winning entries received a $1,000 U.S. Savings Bond. In addition, the schools that those students attend received a plaque commemorating the contest winner. Most importantly, the winning designs will be printed as stickers to be used on light switch plates in all schools of the respective winning level.

Each school had the opportunity to develop the preliminary round of the contest in a manner appropriate for their environment. There were many individuals, throughout the school system who played a part in making this contest happen. Principals were responsible for selecting the methodology used to decide which entry would represent their school in the final judging.

ELEMENTARY SCHOOL WINNER

MIDDLE SCHOOL WINNER

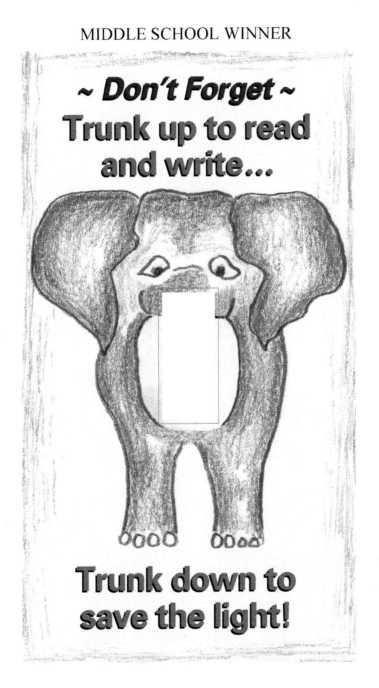

HIGH SCHOOL WINNER

Librarians were the central contact point at each school and disseminated contest information to all students. To promote the contest, school libraries put up extensive displays relative to the contest and the conservation of resources. Science and art teachers told their students about the contest and encouraged students to enter a design. All students learned a little more about energy use, energy costs and energy conservation as they thought up energy-saving slogans and eye-catching designs. By entering this contest, students had the opportunity to promote energy conservation and reap some of the rewards for themselves!

The final judging took place on March 15, 2006. Entries were divided into three categories based on the level of school. There are 12 high schools, 12 middle/intermediate schools and 44 elementary

schools within LCPS. Winners were selected by a panel of seven judges who reviewed the entries and awarded points on four judging criteria: artistry, creativity & originality, overall impression, and their ability to communicate a clear and positive message to turn off the lights in unoccupied areas.

The grand prize winners were announced at the Tuesday, March 28th LCPS School Board Meeting at the LCPS Administrative Building. All three winning students were able to attend with their families. In addition the principals from each of the student's schools were in attendance. The chairman of the school board, Robert F. Dupree, Jr., awarded the prizes to the students, and to each school principal.

During the awards presentation, one potential benefit of the contest was explained: If lights are turned off for one additional hour each school day in a classroom with 16 four-lamp light fixtures, LCPS will avoid spending more than $25 per school year in that classroom. With well over 2,000 classrooms, the potential of avoiding more than $50,000 in expenses each school year exists. Small changes in habits regarding energy use can make a big difference.

Join Us in the Energy Conservation Fight — When You Leave a Room, Turn Out the Light!

The Winners are: High School: Shitij (Dave) Gupta, Broad Run HS Middle/Intermediate: Lindsay Schleifer, Harmony IS Elementary: Ryan Hale, Dominion Trail ES

Index